VOLUME 4

DOUGLAS
DC-6 and DC-7

By Harry Gann

Copyright © 1999 Harry Gann

Published by
Specialty Press Publishers and Wholesalers
11605 Kost Dam Road
North Branch, MN 55056
United States of America
(651) 583-3239

Distributed in the UK and Europe by
Airlife Publishing Ltd.
101 Longden Road
Shrewsbury
SY3 9EB
England

ISBN 1-58007-017-5

All rights reserved. No part of this book may be reproduced or transmitted in any form or by any means, electronic or mechanical including photocopying, recording or by any information storage and retrieval system, without permission from the Publisher in writing.

Material contained in this book is intended for historical and entertainment value only, and is not to be construed as usable for aircraft or component restoration, maintenance or use.

Front Cover photo: Boeing Historical Archive
Back Cover photos: Boeing Historical Archive

Designed by Dennis R. Jenkins

Printed in the United States of America

TABLE OF CONTENTS

THE DOUGLAS DC-6 AND DC-7

FOREWORD 4

CHAPTER 1 — **DOUGLAS TRANSPORTS** 5
FROM THE BEGINNING

CHAPTER 2 — **THE DC-6** 11
A SOLID DESIGN

CHAPTER 3 — **DC-6 VERSIONS** 31
AIR FREIGHTERS AND MILITARY TRANSPORTS

CHAPTER 4 — **THE DC-7** 53
GIVE THE CUSTOMER WHAT HE WANTS ...

COLOR SECTION — **THE DC-6/7 IN COLOR** 65
ALL OVER THE WORLD

CHAPTER 5 — **DC-7 VERSIONS** 69
... BUT DON'T LOSE MONEY

CHAPTER 6 — **EXTERNAL INFLUENCE** 87
THE LOCKHEED CONSTELLATION

CHAPTER 7 — **MISCELLANEA** 89
ACCIDENTS, FREIGHTERS, AND PYLON RACING?

APPENDIX 1 — **DC-7 TRIVIA** 96

APPENDIX 2 — **DC-6 AND DC-7 RECORDS** 96

APPENDIX 3 — **SUMMARY OF PRODUCTION AND DELIVERY DATA** 97

APPENDIX 4 — **SUMMARY OF SELLING PRICES** 97

APPENDIX 5 — **STATION DIAGRAMS** 98

SIGNIFICANT DATES 100
IMPORTANT DATES IN THE HISTORY OF THE DC-6/7

DOUGLAS DC-6/7

FOREWORD

ACKNOWLEDGEMENTS

The Douglas DC-6/7 series, along with the Lockheed Constellation, played a major role in the post-World War II period in reducing the time spent traveling among world population centers. The potential realized by these aircraft in moving goods and people in an ever increasing market signaled an era in aviation technology that spurred other disciplines: electronic communication, banking and other forms of transportation increased levels of service. Airline transportation was the driving factor.

To the Douglas Aircraft Company, the commercial line of transports, coupled with the military aircraft produced, brought worldwide leadership and recognition. There were few airlines which were not operating at least one of the Douglas DC-3, DC-4, DC-6, or DC-7 transports. This record amplified the reputation that the company had acquired during wartime production of almost 40,000 aircraft.

The DC-6/7 series, especially the DC-6B, were acknowledged as some of the finest aircraft designs ever by aviation students and historians. This reputation was gained by economic and performance achievements.

The success of the aircraft was the peak of the Douglas Aircraft Company. In an effort to maximize the economic benefits of the piston-engine product line to the company and its stockholders, Douglas "drug its feet" in getting into the race to supply the market with turbine-powered transports. When a decision to enter was made, the company did not offer a wide enough product line. In contrast, Boeing in the 1950s had to develop new product or leave the business. Boeing took up the challenge, made fewer mistakes, and was victorious before the century's end.

Many individuals have contributed to this story. Pat McGinnis at Boeing in Long Beach allowed me free access to the Douglas files. Retired Douglas personnel that were of great assistance were test pilots Bill Smith and Len Burke. Bill Smith was slated to be the pilot of the DC-7B that crashed as a result of the mid-air collision with the Northrop F-89 but was pulled off at the last minute to participate in another project. As an Air Force pilot, Len Burke flew both DC-6 and Constellation aircraft as part of the presidential fleet. Both were able to guide this non-pilot in his verbiage. Walt Bohl, a retired United Airlines captain and fellow AAHS board member, also critiqued the text.

Other AAHS associates that contributed were Tim Williams, Al Hansen, Byron Calomiris, and John Dzurica.

Milo Peltzer supplied facts regarding the use of DC-6 and DC-7 aircraft as air tankers while Carla Wallace and William Friedman of the National Oceanic and Atmospheric Administration provided information on the "Hurricane Hunters."

I used many sources for references. I would like to single out the Air-Britain publication, *The Douglas DC-6 and DC-7 Series*, by John A. Whittle as a prime source.

My son, Russell, helped me when my computer got the best of me.

All of their assistance is greatly appreciated.

Any mistakes made in this story are solely the fault of the author.

Harry Gann
1999

Compania Mexicana de Aviacion (CMA) through it association with Pan American was, at one time, the largest airline in Latin America. It operated this DC-7C out of Mexico City to the U.S. and to other Latin American countries. (Boeing Historical Archive)

Douglas Transports

From the Beginning — Chapter 1

The Douglas reputation for building aircraft, especially airliners, is renowned around the world. The DC-6/DC-7, as well as the esteemed DC-3, is acknowledged as the ultimate design in prop-driven transport aircraft. This reputation was built on a solid background of conservative and innovative engineering. The first aircraft built with the Douglas label, the Cloudster, was engineered for Mr. David R. Davis in 1921 for the single purpose of being the first aircraft to fly nonstop across the United States. It did not accomplish this when the Liberty engine failure forced an emergency landing in Texas, but it nevertheless set aviation records. Later, it was converted to a ten-passenger transport after Davis lost interest in the transcontinental flight. Claude Ryan's airline operated it between Los Angeles and San Diego, California. The Cloudster was the first of over 50,000 aircraft to carry the Douglas name.

After Davis left the Davis-Douglas Company, Donald W. Douglas was determine to carry on and he was successful in raising sufficient financial backing to enter several competitions for new aircraft for the U.S. Navy and the fledgling U.S. Air Service. A winning proposal was written for the C-1 transport – a Liberty-powered biplane. The basic requirement for this logistic freighter was that it could transport a Liberty engine as cargo. Douglas decided to qualify this aircraft for a commercial aircraft type certificate (ATC), but that market was not present and only 26 military versions were built. Still, lessons learned on this aircraft were to serve the company well in later endeavors.

The first Douglas-built commercial transport was the Mailplane. Based on the design of the O-2 observation aircraft used by the Air Service, the Mailplanes help to inaugurate Western Air Express (WAE) airmail, passenger, and cargo service from Los Angeles to Salt Lake City, Utah in 1926. The U.S. Post Office was the largest buyer of the 59

Of the more than 50,000 aircraft that were built carrying the Douglas name, approximately 20,000 were transports. The Liberty-powered biplane Cloudster became the first, making its initial flight in 1921. After an attempt at becoming the first aircraft to make a nonstop transcontinental flight, it was converted to a ten-passenger transport. (Boeing Historical Archive)

Although the C-1 was designed and built to satisfy a requirement from the U.S. Army Air Service for a logistic transport, Douglas made an unsuccessful effort to interest airlines in its use. (Boeing Historical Archive)

Mailplanes constructed at the Douglas facility on Wilshire Boulevard in Santa Monica, California, during 1925 and 1926. When the Post Office relinquished the mail transportation to private operators in 1926, the Mailplanes were used by the airlines that were organized to pick up the demand for delivering the airmail. If an occasional passenger was carried, this was an extra bonus. The Mailplane secured its place in the history of aviation transportation. One marked in WAEs colors is on display at the National Air and Space Museum (NASM) in Washington, D.C.

In 1928, Douglas felt that the market might be right for a plush, twin-engine, amphibian aircraft. Originally named the Sinbad, the design was renamed Dolphin and aircraft were sold to the military, airlines, and private operators. The Wrigley family used these aircraft to carry passengers between the Los Angeles area mainland and their resort on Catalina Island. Of the 59 aircraft produced, only three were used in airline operation. Clearly the Dolphin design was not suitable for mass air transportation, however the Douglas experience bank had another deposit.

In 1933, Transcontinental and Western Airlines (TWA) approached Douglas as well as four other American aircraft companies to determine if they would be interested in submitting a proposal for a three-engine transport to replace their wooden-wing and steel-tube fuselage Fokker F-10 transports. After much thought, Douglas decided to submit a twin engine all metal design instead. This concept used new Pratt & Whitney or Wright air-cooled radial engines that could develop 700 horsepower for take-off. The resultant aircraft, the DC-1, was deemed the competition winner by TWA. Only one DC-1 was built, but that aircraft design literally changed history.

The production version, the DC-2, won an order of 20 aircraft from TWA. When the engine manufacturers were able to increase the power available, Douglas stretched the DC-1 fuselage the equivalent of one row of seats in the production version. Many U.S. and foreign airlines followed TWA in procuring the DC-2. This basic configuration was again stretched and widened to meet an American Airlines requirement, and the DC-3 was created. Douglas

The biwing Mailplanes were the first Douglas aircraft that were specifically produced as transports. The primary customer was the U.S. Post Office, which ordered them to replace the World War I surplus DeHavilland DH-4s as airmail carriers. Western Air Express also procured them to carry the mail, as well as passengers and other cargo, on their route from Los Angeles to Salt Lake City. An M-4 Mailplane is shown with a DC-6B. (Boeing Historical Archive)

While only one aircraft was produced, the DC-1 was a major factor in directing the explosion of aviation transportation. (Boeing Historical Archive)

learned early in its design experience the advantage of engineering-in the ability to stretch the fuselage to accommodate future demands of the market. This pre-thought was to become a major factor in the selling of commercial air transports. Some manufacturing companies were to learn this lesson the hard way.

The twin-engine high-wing DC-5 was developed by Douglas in 1939 to complement its available range of aircraft for the airline market. The objective of the DC-5 was to serve the smaller commuter-type airlines. This concept proved to be premature with the entry of the U.S. into World War II, and only 12 aircraft

A DC-2 (foreground) flies formation with its more famous derivative, the DC-3. The DC-2 was the production version of the DC-1, and 130 of these 14-passenger transports equipped most of the U.S. airlines as well as many overseas operators. The follow-on was initially produced as a sleeper version, but the dayplane DC-3 rapidly became the prime aviation transportation tool. By 1938, the DC-2 and the DC-3 carried 80% of the American aviation transportation load. (Douglas Historical Foundation via Harry Gann)

DC-5 production was limited to 12 aircraft mainly because of the outbreak of World War II, which curtailed the demand for a short-range, commuter-type aircraft. (Boeing Historical Archive)

were built at the Douglas El Segundo facility. Of these, eight were destined for military usage. Much later, the jet-powered DC-9 would successfully fill the niche in the commuter transportation field market prematurely targeted by the DC-5.

Requirements during World War II resulted in the production of over 10,000 DC-3s which were called the C-47/R4D/C-53/Skytrain/Dakota by the military users. Their high availability and low initial cost became a factor in the world's transportation industry when many airlines that were started after the war began with the civilized C-47s.

In an effort to respond to the postwar market for a "DC-3 replacement," the Douglas advanced design group analyzed various configurations. However, none of these studies appeared to produce an economically viable product. Consequently, Douglas went with an approach that took advantage of the large quantity of DC-3 airframes available and offered the DC-3S or Super DC-3. Douglas proposed that a customer would send in his DC-3 or C-47 to be modified by stretching the fuselage 80 inches, attaching new wings and empennage, along with uprated Wright engines that developed 1,450 horsepower. This would greatly reduce the cost from that of a completely new aircraft.

However, only 105 aircraft were modified to the DC-3S configuration, mostly for the U.S. Navy and Marine Corps. Only Capital Air Lines was to take advantage of this exchange plan when it ordered three of the modified DC-3s. Convair was to capture most of the DC-3 replacement market with its 240/340/440 series of twin reciprocating engine aircraft.

In an effort to extend the life of the many wartime-built DC-3 series aircraft, Douglas developed the DC-3S, which was a major modification of the "Gooney Bird." The DC-3S did not become a viable sales item, while the DC-3 continues to be used to this day. (Douglas Historical Foundation via Harry Gann)

The Early Douglas Four-Engine Transports

The world's airlines, pushed by the expansion of international trade, needed to fly farther with greater capacities. Non-stop transcontinental service was an immediate goal of the U.S. airlines, with trans-oceanic flying also being desirable. To accomplish this, larger aircraft with increased power were needed. The success of the DC-3 helped to kindle an interest in this travel growth, and Douglas was to take a leading step in the planning for the inevitable expansion. Even before DC-3 service was inaugurated in 1936, studies were being undertaken to determine how this growth could be attained.

In 1935, United Air Lines president W. A. Patterson suggested to Douglas that it was time to consider the production of an aircraft to meet the new market. He envisioned an aircraft that would be larger and more luxurious than any other mode of travel. Donald W. Douglas agreed, but pointed out that the cost of developing this transportation milestone would be beyond the capability of a single company. He further suggested that this might be accomplished with the aid of an airline consortium where the cost could be shared by several companies. Patterson resolved to accomplish this task and called a conference of interested airlines to develop a set of specifications acceptable to all parties.

On March 23, 1936, an agreement was signed among the Douglas Aircraft Company and United Airlines, Eastern Air Lines, American Airlines, Transcontinental Western Airlines, and Pan American World Airways to build an aircraft that would meet their joint requirements. Each airline was to commit $100,000 for the design and construction of a prototype, to be designated as the DC-4. An initial production run of 20 aircraft was envisioned.

The DC-4 was to be capable of carrying 42 passengers with a crew of five with such amenities as sleeper berths, dressing and smoking rooms, and a fully equipped galley. The 65,000-pound aircraft would have a range of 2,000 miles at a cruising speed of 190 mph. An air-conditioned and pressurized cabin was considered, along with many other advanced features. These features proved to be too advanced, and Douglas decided to abandoned the DC-4 after the production of one aircraft and extensive route testing by United Airlines. The company would start over with a new aircraft with reduced capability. This aircraft would also be called the DC-4, the previous aircraft being redesignated DC-4E. Subsequently, Pan Am and TWA decided to drop out of the consortium in order to support Lockheed and its Constellation design.

A production run of 24 DC-4 aircraft was assigned to the Douglas facility at Santa Monica, California, for delivery to United and American Airlines. However, before the airlines could obtain delivery, World War II

Douglas efforts to develop a longer-range passenger airplane (DC-4, nee DC-4E) were aided by a consortium of five airlines, although the initial effort was ultimately unsuccessful. Many lessons were learned in this joint project and the initial product was scaled back in the resulting production version. (Boeing Historical Archive)

Wartime production priorities interrupted the commercial production of the DC-4; however, many of those built as military aircraft would be used by the world's airlines after the cessation of hostilities. (Boeing Historical Archive)

began and these aircraft were reassigned for delivery to the U.S. Army Air Corps and the U.S. Navy as C-54 and R5D aircraft, respectively. The government initially pressed Douglas to concentrate production facilities on the proven DC-3/C-47 and abandon further development of the four engine-DC-4s. Douglas was successful in convincing the government of the utility of the DC-4 as a military vehicle and that the company could meet wartime production rates for the C-54 and the C-47. The C-54/R5D line continued at Santa Monica and at a new factory in Chicago, and 1,162 of these four-engine transports were built during the war. The airlines would have to wait until the war was settled in Europe and the Pacific before they would take delivery of the DC-4s. After the war, 79 additional aircraft were built for commercial operators.

The DC-4E was 150 percent bigger than the DC-3. Route testing by United Airlines led to a re-evaluation of the overly ambitious project. (Boeing Historical Archive)

The DC-6

A Solid Design — Chapter 2

Although outwardly similar to the DC-4, there were many changes and improvements in the DC-6. The main visual differences were the longer fuselage, a squared-off vertical tail, square passenger windows, a lack of de-icer boots and longer engine nacelles. Design features also included new engines, cabin pressurization, complete air conditioning, thermal anti-icing of the windshield, wing and empennage surfaces, an electronic autopilot system, double-slotted flaps, new brakes, and electrically operated cowl flaps.

The wing design was retained, although the aluminum material was changed from 24ST to the new 75ST which lowered the wing weight by 800 pounds. The fuselage length was increased 81 inches and the gross takeoff weight was increased from 73,000 to 92,000 pounds. The aircraft was capable of cruising at speeds in excess of 300 mph, and at maximum gross weight, and could maintain flight with only two engines in operation. The range capability made both trans-continental or trans-oceanic operations feasible.

Douglas made claims to many "firsts" in the DC-6 design, including:

- First pressurized airliner whose fuselage could be maintained at comfortable temperatures at all altitudes.
- First airliner with a completely automatic cabin pressure control system.
- First airliner that was completely air-conditioned (both temperature and humidity).
- First airliner to use radiant-heated floors and walls guaranteeing even temperature distribution at all times.
- First airliner fully air conditioned on the ground as in the air.
- First airliner with electric propeller de-icing.
- First airliner to utilize 75S, the then-new miracle metal aluminum alloy, to increase strength and save weight.
- First airliner with preloaded cargo containers and facilities for handling them built into the airplane.
- First airliner with fog-proof and frost-proof passenger windows.

This combination mock-up shows both the R-2800 engine of the XC-112 and the Allison V-1710 engine of the XC-114. The XC-112 became the YC-112A and one aircraft was produced, while the XC-114 concept did not result in aircraft being built. This photograph was taken September 13, 1945, just 11 days after the end of World War II. (Boeing Historical Archive)

Sans markings and identification numbers, the YC-112A flies parallel to the California coast just north of the Douglas Santa Monica facility. This test flight was made on March 13, 1946, shortly after first flight of February 15. (Boeing Historical Archive)

Cabin

The fuselage was an all-metal semi-monocoque structure consisting basically of three sections. The cockpit extended from the nose to station 129; the cabin and cargo section extended from station 129 to station 938; and the aft section, which included the horizontal and vertical stabilizer stubs and an attachment for the tail cone, extended to station 1115. The three sections were bolted together at butt angle joints to form an integrated fuselage.

The flight compartment featured a clear-vision windscreen designed to withstand the impact of a four-pound bird at 350 mph. This windshield was maintained in an ice-free condition by the passage of hot air between its outer and inner panes. The outer pane was 1/4-inch full-tempered glass and the inner pane was a laminate of two layers of 3/16-inch glass with 1/4-inch vinyl filler. The nose section provided support for the nose landing gear and nose well doors and contained the well for the wheels.

The fuselage center section may be considered a two-story structure. The upper "story" contained the main cabin area, including passenger seating and sleeping accommodations, the buffets, and the luxury lounges.

The YC-112A did not serve long with the Air Force, which released it for civilian usage. Mercer Airlines was the last operator. (Harry Gann)

Also included was a forward cargo compartment, extending from station 149 to 260.5, that was designed for a cargo density of 150 pounds per square foot (50 pounds per running inch). The lower "story" contained the main cargo compartments and specialized equipment such as air-conditioning and hydraulics.

The main cabin support structure consisted of four lines of fore-and-aft beams that used extruded magnesium channel sections. Cross crosswise with 75ST extruded hat sections which were spaced at 40-inch intervals to support the seats. The flooring was designed to accommodate high distributed loadings as well as high loads applied over a small areas, such as a woman's weight concentrated in the pressure of her high-heel, without damage.

The passenger windows were hermetically sealed, fog-proof and frost-proof, and were designed to withstand pressure loads of 8.33 psi ward and was hinged at the forward jamb, leaving a clear opening of 36x72 inches. The forward flight compartment door also opened outward and had a clear entry of 30x60 inches.

The forward lower cargo compartment extended from station 149.6 to 341: the aft cargo compartment extended from station 600 to 760. The doors of these compartments opened outward and were hinged to swing downward. Both doors were 37 inches long by 28

The prototype DC-6 made its first flight July 10, 1946, and was used in the flight test and certification program prior to delivery to American Airlines. The passenger windows were rectangular double-panel Plexiglas with an air gap to prevent frost. (Boeing Historical Archive)

bulkheads and beams supported the fore-and-aft beams. The floor beam structure was designed to withstand, in addition to cabin floor loads, the cargo basket loads that could be carried on overhead transverse monorails and two longitudinal rails.

The flooring in the passenger section was flat dural 75ST sheet with hat sections of 24ST, running (more than a ton per window). The 16x18-inch windows consisted of a Plexiglas outer pane, 5/16-inch thick, and an inner pane 1/8-inch thick with a 1.8-inch gap between the panes. There were six auxiliary exits, measuring 21x26 inches, which were hinged at the forward edge and arranged to open outward.

The main cabin door opened out- inches clear vertical in size, or 45 inches measured on the fuselage circumference. These compartments were designed so that the cargo carriers or baskets could be preloaded (unique for that time) and carried aboard for flight by means of an overhead monorail system, saving considerable stopover time. Each compartment held four baskets that were 36 inches wide and had a volume of 30 cubic feet. Each basket could carry 500 pounds.

The lower cargo compartment floor structure was strong enough to carry maximum loads of 25 pounds per cubic foot, or a maximum of 26.6 pounds per running inch. It was possible to carry a cargo load of 2,842 pounds in the forward compartment and 2,394 pounds in the aft compartment.

The fuselage was pressurized from the forward bulkhead at station 64 and the slant bulkhead over the nose wheel, to the pressure dome at station 938. The design permitted a pressure differential of 8.33 psi, but was normally operated at 4.16 psi, which was the pressure differential corresponding to 8,000 feet altitude at 20,000 feet. On the DC-6B, the normal cabin pressure differential was raised to 5.46 psi, which was the equivalent of 8,000 feet while cruising at 25,000 feet.

The buffet was divided into two sections, one on each side of the entrance door, to allow quick loading of food and convenience during serving. The space directly in front of the entrance door became the flight attendant's workroom with seat and desk when the airplane was in flight. The seat had a safety belt for take-off and landing and both the seat and desk folded out of the way when not in use.

The general cabin lighting was semi-indirect with the fixtures installed on both sides of the cabin center dome and running the entire length of the cabin. The amount of electrical power used for the DC-6 cabin lighting was ten times that used in the DC-3.

The cabin interior could be maintained to within 3°F of any desired temperature. The air entered through an opening in the leading edge of each wing, was ducted

Douglas Aircraft frequently used professional models in publicity photos. Norma Jean Baker, better known as Marilyn Monroe, is shown seated inboard in the last row. This photograph demonstrates the four-abreast, day configuration of the DC-6 family. (Boeing Historical Archive)

through supercharger blowers, and transferred through an after-cooler or a heater, as effected by the cabin temperature controls. The DC-6 floor and side panels were heated by circulating warm air through ducts which distributed the air up beyond the passenger widows to outlet grills. Thus, the floor itself was heated and the walls of the fuselage are heated at each seat. Individual air outlets supplying cool outside air were provided adjacent to each pair of seats and in the lounges and crew quarters. These outlets permitted occupants to adjust local ventilating conditions to individual taste.

The DC-6 embodied three features to ensure a minimum of noise to distract the passengers. A sheet of laminated mica, varying from

The DC-6 had various sleeper berth options. The berth locations could be forward or aft, or the entire cabin length. Shown is a set of sleeper berths, one configured for day and one made up for sleeping. Model Norma Jean Baker (Marilyn Monroe) demonstrates the use of the lower berth in a DC-6. (Boeing Historical Archive)

DC-4 and DC-6 aircraft are shown during final preparation for delivery to their customers at Clover Field during July 1947. In the foreground is the VC-118 Independence *(left) parked nose-to-nose with the VC-54C* Sacred Cow, *both presidential aircraft. (Boeing Historical Archive)*

0.010 to 0.020 inches in thickness, was cemented directly to the airplane skin generally throughout the fuselage. This acted as a sound deadener by absorbing vibration through dampening between the mica laminations. In addition, blankets of fiberglass were cemented together with a non-porous film and mounted over the mica, covering both the spaces between the frames and over them. The engine exhaust outlets were directed away from the fuselage to reduce the noise to the passengers and to partially hide the flames during high power settings.

The DC-6A emergency egress was provided by either five or seven window escape hatches, while the DC-6B had nine emergency window escape hatches.

Crew Accommodations

The size of the DC-6 series crew varied with the airline usage. Overland versions usually consisted of the pilot, co-pilot, flight engineer, and two flight attendants. Overwater configurations added a navigator and radio operator in the area just behind the fight compartment usually provided for forward cargo.

Tail cones waiting their turn to be mated to DC-6 aft fuselages. The tail cones were attached by bolts at station 938, the rear pressure dome. (Boeing Historical Archive)

The windshield was dual-panel, bird resistant, and incorporated heated anti-icing. The DC-6 nose section extended from station −3 to 129. (Boeing Historical Archive)

DC-6 Passenger Seats

During the immediate postwar time period, Douglas chose to design and fabricate in-house many of the components used in its aircraft, including the passenger seats and sleeper chairs. These chairs fitted the contour of the airplane at the outboard side, giving an extra wide seat while still allowing a 21-inch aisle between arm rests and 26 inches at seat back hinges. The seats could be reclined smoothly by a Douglas-designed hydraulic mechanism. The seat was designed such that when the aircraft was vacated, all of the seat backs slowly moved back to their normal position. The

American Airlines was the first airline to order the DC-6 aircraft and was to eventually receive 50. As indicated by the location of the smaller berth windows on the fuselage, this model could accommodate 50 passengers as a dayplane or 24 in sleeper berths. (Boeing Historical Archive)

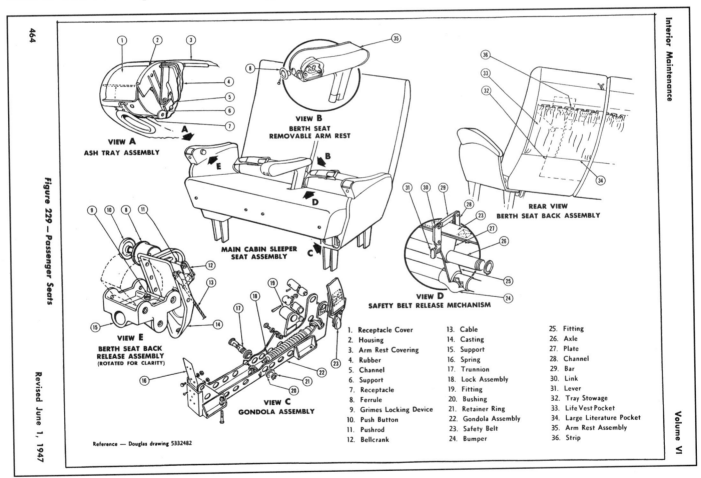

Douglas Aircraft designed and built the passenger seats for most of the DC-6 and DC-7 aircraft. The seats were approximately 20 inches wide and pitched at 40 inches. The seat backs could be quickly lowered to convert to sleeper berths. (Boeing Historical Archive)

seats were also designed such that no center support was tied to the floor, leaving full width 10-inch high space for luggage or foot room under each seat.

The sleeper chair was designed to be made into a berth within 30 seconds. The seats had a cantilever sliding track mechanism in the leg tie-downs that released the seat bottoms to a horizontal position. Alternate seat backs let down to a horizontal position while the other seat backs slid to a vertical position and became partitions between the berths. The center armrest of each seat was then removed and stowed under the seat. When made up, the seats became a 42x76-inch lower berth.

The upper berth was concealed in the ceiling of the fuselage and with one operation could be swung down and made into a berth within 10 seconds. Both upper and lower berth curtains and end panels fell into position when the upper berth was lowered. The upper berth was ready for occupancy when it was lowered.

DC-6 Cargo Provisions

The DC-6 was one of the first aircraft to have preloaded cargo-container handling facilities within the airplane. There were two cargo compartments fore and aft within the belly of the aircraft fuselage, eliminating the necessity for external cargo containers. The forward compartment had a capacity of 2,842 pounds and the aft compartment a capacity of 2,394 pounds. The forward compartment contained 203 cubic feet of space and the aft 171 cubic feet when not using the preloaded containers.

Both of these cargo compartments contained facilities for handling preloaded cargo containers. These facilities consisted of longitudinal rails attached to the aircraft floor beams that supported lateral rails to which the cargo baskets attached. The 37x45-inch doors to these compartments hinged outward and downward, thus dropping the door out of the way during loading operations. Each cargo

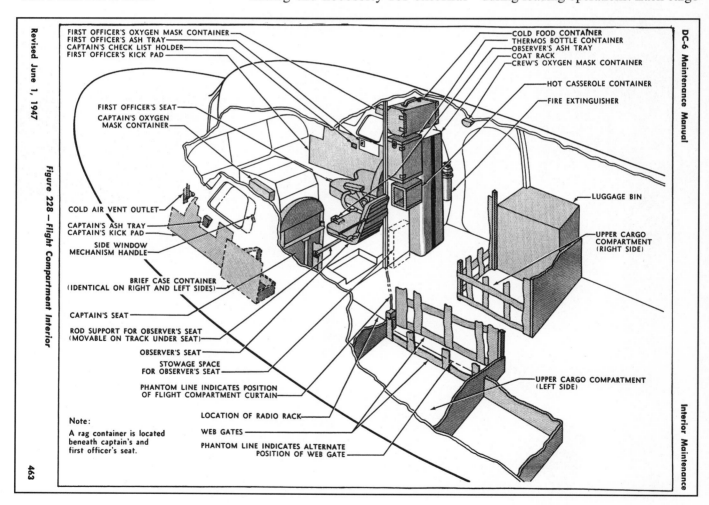

Shown is the overland configuration of the DC-6 flight compartment. For overwater flights, radio operator and navigator positions replaced the upper cargo compartments. (Boeing Historical Archive)

These photographs were taken in mid-1946 and illustrate an early configuration of the DC-6 flight compartment. (Boeing Historical Archive)

compartment could accommodate four baskets, which were loaded into the airplane according to the sequence of destination.

External loading equipment, incorporating an overhead monorail, lifted the baskets and transferred them onto the lateral rails which then rolled forward or aft along the longitudinal rails within the airplane. Mechanical loading devices permitted the unloading and reloading of four baskets within a 10-minute stopover period without interference with other ramp service activities or passenger movement.

DC-6 Wings

The DC-6 wing structure was similar in design to the DC-4 and was the same size. The main improvement to the wing was the use of mostly 75S aluminum in the structural members and skin in place

Delta purchased seven DC-6 aircraft, which were dayplanes. The first six were configured to the basic Detail Specification 477B, and the last to Type Specification 1156. (Boeing Historical Archive)

of the 24S used in the DC-4. Higher strength was gained with the use of this type of aluminum alloy. The wing section at the root was NACA 23015 tapering to NACA 23012 at the tip, as were all of the DC-6/DC-7 series. (The last two digits indicate the thickness percentage.)

The full cantilever, all-metal wing was composed of a center section approximately 70 feet in span, with a detachable outer panel and wingtip on each side. The center section also included the four engine nacelles and was assembled integrally with the center section of the fuselage. The total wing area was 1,463 square feet with an aspect ratio of 9.44.

Three transverse spars bridged the center section structure, the front and center spars continuing onto the outer panel. Rib and bulkheads, and the metal covering stiffened with hat-type stringers, characterized both center and outer panel structure. The entire wing structure, including double-slotted flaps and

The Pan American Grace Airways DC-6 configuration was built to the basic Detail Specification 477B and was a convertible day/sleeper version. In a static condition, the vertical tail rose to 28 feet 5 inches. (Boeing Historical Archive)

the ailerons, accounted for 16% of the entire airplane empty weight.

The main landing gear support structures were located in the inboard nacelles with the main shock strut fittings on the center spar and the forward brace fittings on the front spar. These fittings were forged of 14ST material. The center spar landing gear fitting, because of its extraordinary strength, incorporated the airplane hoist fitting, which was designed to sustain the aircraft's gross weight. Located on the rear spar inboard of the inboard nacelles were the airplane jacking points.

The wing outer panels were bolted to the wing center section on each side and were removable. The wing tips were bolted to the outer panels and were also removable. An aileron was attached to each outer panel; the right aileron incorporated the trim tab, the left aileron a divided spring control tab. Two hydraulically operated wing flaps were designed to move aft and down on linkages attached to the wing center section.

The aileron was of conventional type, and of all-metal, no-spar construction with flush-riveted covering. The weight of each aileron was 253 pounds.

DC-6 Empennage

The cantilever, fixed horizontal and vertical stabilizers were two-spar construction and were bolted to stubs on the aft fuselage section. The smooth skin covering was reinforced by pressed ribs of two-piece composition. No spanwise stringers were used in these structures which featured 75S spar caps and 24STAL skin.

The leading edge sections incorporated heat anti-icing passageways and were detachable units. The rear spars were stepped extrusions with the attachment fittings machined integrally in the large stepped end.

The installation of the observer's glass dome on the upper forward fuselage indicates that this Philippine Air Lines DC-6 carried a navigator and was an overwater version. The number of sleeper windows suggests that the passenger configuration was the standard 50 day/24 sleeper berths. (Boeing Historical Archive)

The continuation of the caps across the tail section also was made by integral fittings that were machined from the standard extruded section.

The elevator structure consisted of pressed ribs with metal, flush-riveted covering. The smooth skin covering was reinforced by pressed ribs of two-piece composition. No spanwise stringers were used in these structures, which featured 75S spar caps and 24STAL skin. Each elevator weighed 215 pounds. Conventional cloth-covered metal rib and spar construction was used for the rudder, which weighed 120 pounds.

The tail cone was attached to the fuselage tail section with bolts. The weight of the complete tail section was 1,406 pounds.

Royal Dutch Airlines, better known as KLM, purchased eight of the basic DC-6 aircraft. The position of the small windows in the aft fuselage indicates that this configuration is a rear sleeper version. Registration shown is in error. The correct registration was PH-TPI. (Boeing Historical Archive)

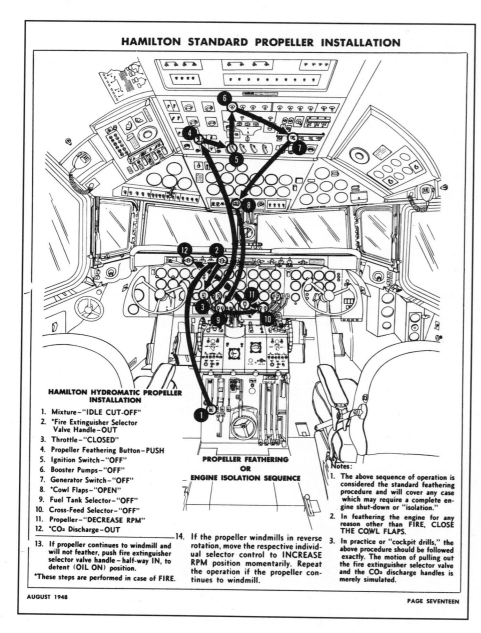

The procedure for feathering the DC-6's Hamilton Standard propeller is shown. (Boeing Historical Archive)

given to the engineers on the DC-6 was to reduce the fire hazards and to ease the maintenance load. This was especially evident in the engine installation design.

The DC-6 was designed with a demountable power plant that incorporated a complete oil system, including the tank, forward of the firewall. Thus, the removal or installation of the engine could be accomplished in a minimum time period and the replacement power plant could be installed in complete working order so that no adjustments were necessary after placing on the aircraft.

Either Curtiss electric reversing propellers or Hamilton Standard 13-feet 1-inch diameter, Hydromatic full feathering, reversible propellers were used on the DC-6, allowing greater ground maneuverability as well as providing a means of reducing the ground roll after touchdown.

DC-6A and DC-6B Engines and Propellers

The DC-6A and DC-6B used either R-2800 CB-16 or CB-17 engines. The CB-17 engine was similar to the CB-16 engine except that higher power settings could be utilized when a higher-octane fuel (Grade 108/135) was used.

The DC-6A/B was equipped with 13-foot 1-inch Hamilton Standard Hydromatic propellers. This was a full-feathering, reversing pitch, constant-speed type, with electric de-icing. Some customers opted for spinners and engine cowling liners that provided a lower cylinder head temperature.

DC-6 Engines and Propellers

A major improvement of the DC-6 was the selection of the Pratt & Whitney R-2800 Double Wasp engines to replace the P&W R-2000 Twin Wasp used on the DC-4. The R-2800 was a well-proven engine used by many wartime aircraft such as the P-47 Thunderbolt, the F4U Corsair, and the F6F Hellcat. This was an 18-cylinder, 2-row, radial, air-cooled engine with a single-stage, two-speed integral supercharger. The DC-6 initially used the CA-15 version that developed 2,100 horsepower at takeoff. Except for minor installation changes, the engines were interchangeable for all positions. The DC-6 also used the CA-18, the CB-16, the –34, and the –83A versions in some later deliveries. The CB-16 engine had a modified supercharger section to incorporate an enlarged impeller and a lengthened impeller inducer. The –34 and –83A engines had a different location of the fuel transfer line mount pad and the oil breather fittings.

One of the prime design criteria

Engine Ratings

The various P&W R-2800 engine versions were rated at basically 2,100 brake horsepower for two minutes at take-off, or 2,400 bhp for two minutes wet. The CB-17 used in some DC-6A and DC-6B aircraft was rated at 2,500 bhp (wet) for take-off. There was some variation in the performance of these engines at altitude depending on the version.

Fuel System

The fuel system installed in the DC-6 could be one of two different arrangements depending upon the preference of the operator. These systems were referred to as the eight-tank and the 10-tank systems. The eight-tank arrangement could accept 3,322 U.S. gallons of fuel, while the 10-tank configuration could carry up to 4,722 U.S. gallons, depending on the exact configuration. Six tanks in the fuel system were incorporated with the structure of the wing between the front and center spar. Collapsible fuel cells supplemented the integral fuel tanks to provide additional capacity for both the eight- and 10-tank fuel systems.

This all-wing fuel installation helped to counteract the gust loads on the wings in addition to providing a safety measure by locating the flammable fuel outside the passenger area.

The fuel system was designed so each engine had its own complete fuel system independent of the other engines with the exception of the cross-feed manifold line which would let the crew direct fuel from any tank to any engine. The crew could then balance the fuel load as desired or use all the fuel available in the event of failure of an engine. Engine driven fuel pumps, and an electrically driven boost pump for reach wing tank, were provided with each engine fuel system. The four individual priming systems were electrically operated.

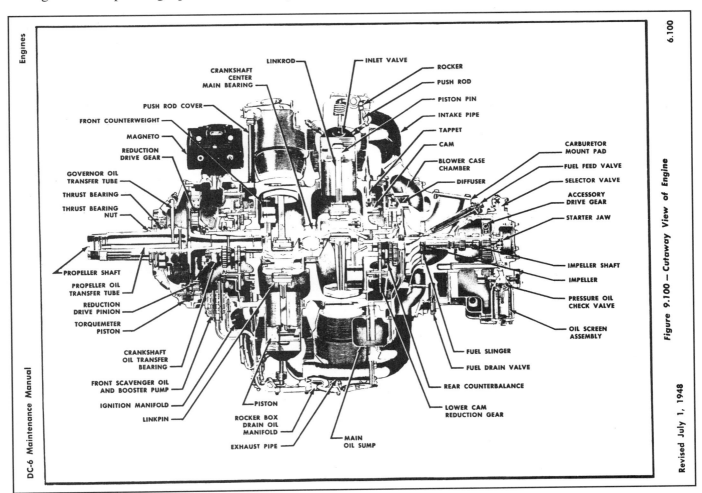

The DC-6 had the advantage of using one of the most efficient piston engines ever produced – the Pratt & Whitney R-2800 Double Wasp. The 18 cylinder, twin row, radial, air-cooled, geared supercharged power plant produced one horsepower per pound of engine weight. (Boeing Historical Archive)

Dump valves, with retractable chutes below the wing tanks, were provided to release fuel in an emergency. The fuel dumping system was arranged such that the crew could select tanks from which fuel was to be dropped and also controlled the quantity of fuel they desired to release.

The basic controls were located in the flight compartment of the airplane for fuel system management. They included four fuel tank selector valve controls for the eight-tank system and six fuel tank selector valve controls for the 10-tank system; two cross-feed valve controls; eight or 10 fuel booster pump switches; and four fuel dump valve controls. Fuel flow meters were installed in the metered fuel line at the engine, between the carburetor outlet and fuel discharge nozzle.

On DC-6A and DC-6B aircraft, the customer could choose from 11 fuel tank options within the eight- or 10-tank arrangements. Total capacity could vary from 3,992 to 5,512-U.S. gallons of fuel.

DC-6 Landing Gear

The DC-6 landing gear was similar to the DC-4, although a number of refinements were incorporated, including an additional safety mechanism which prevented the landing gear from being retracted while any weight remained on the gear.

The landing gear, provided with normal hydraulic and auxiliary extending mechanisms, was capable of extending and locking in the down position by gravity and air loads only. The landing gear control lever, located on the control pedestal, operated both the nose and main landing gears.

The nose gear was steerable hydraulically, which permitted it to turn to an angle sufficient to pivot about one pair of main wheels. This resulted in the aircraft rotating in a small area, easing ground handling.

The main gear incorporated dual wheels and tires on the shock strut. This added greatly to the safety of the airplane in the event of either a

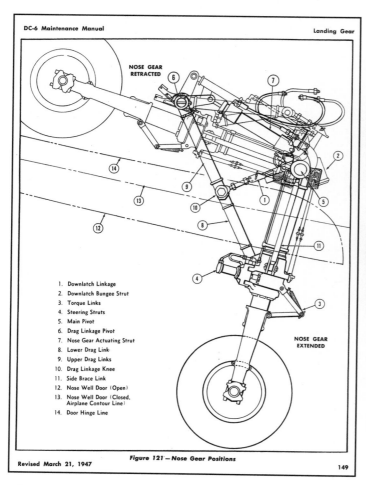

The DC-6 nose gear was steerable and could pivot the aircraft on either of its two main gears. (Boeing Historical Archive)

The DC-6 main landing gear was very similar to the DC-4 gear with some additional safety features incorporated. The tires were inflated to 95 psi. (Boeing Historical Archive)

tire blowout or brake failure. If a tire failed, the plane would not drop on a flat rim since the remaining tire on that side of the aircraft would sustain the weight. Likewise, if there was a failure of the brake in any one wheel, the remaining brakes were still operable. If there was a hydraulic failure, the brakes could be applied with air as an emergency means to stop the airplane.

A faired, non-retractable tail skid, supported by a shock strut, protected the fuselage tail section from possible damage in the event of a tail strike during landing, or an over-rotation during take-off.

DC-6 Flight Control System

Most flight controls were duplicated at the pilot's and co-pilot's station, and either pilot had convenient access to all control wheels, trim wheels rudder and brake pedal, engine control levers, flaps lever, and landing gear lever as well as radio and radio navigation controls. The moveable flight control surfaces were operated through conventional two-way closed cable systems except for the wing flaps, which, although cable controlled, were operated hydraulically. Automatic pilot servos units were attached in parallel to the aileron, elevator, and rudder cable systems.

Over 300 miles of high tensile steel wire was used to fabricate the control cables. More than 500 pulleys, control arms, and rotating shafts

The landing flaps were hydraulically actuated and mechanically synchronized to prevent the flap on one side of the airplane from either going down farther or not as far as the flap on the other side. This was accomplished by connecting the flaps together with a cable. (Boeing Historical Archive)

moved on sealed and lubricated ball bearings. All flight control surfaces were metal covered, except for the rudder, which was fabric covered.

Trim tab control surfaces were provided on the right aileron, on both elevators, and on the rudder. Spring control tab surfaces were installed on the left aileron, on both elevators, and on the rudder (a single tab on the rudder acted as both a trim tab and a spring control tab). Spring control tabs were operated with the main flight controls, providing aerodynamic boost for the control surfaces and thus reducing operating loads.

The employment of an aerodynamically assisted rudder "flying tab" afforded flight control loads without hydraulic or electrical boost systems.

A cable-operated control-surface lock system, manually controlled from the flight compartment, was provided to mechanically secure the ailerons, elevators, and rudder in the neutral position when the airplane was not being operated. This locking system also prevented advancing power to all engines when the controls were locked. However, one engine could be run up at a time for a check.

The DC-6 was equipped with double-slotted, hydraulically operated flaps of the type initially used in the A-26 Invader. This type flap was provided with an airflow deflector

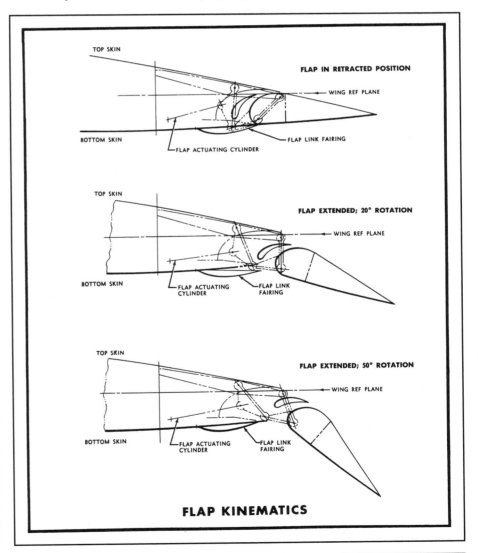

FLAP KINEMATICS

between the wing and the flap that moved with the flap. These deflectors directed the airflow to the upper surface of the flap and provided additional lift, enabling the airplane to take off and land at lower speeds.

DC-6 Passenger Configurations

The passenger cabin was 7 feet high, 8 feet ten inches wide at floor level, and was divided into two sections by the buffet and coatroom. The passenger capacity of the DC-6 varied from 42 to 80 while the DC-6B could carry from 56 to 102 passengers depending on the airline's choice of configuration.

In the sleeper version (DC-6 only), the forward section had numerous seating arrangements of double seats that could be made into lower berths. The aft section also had numerous seating arrangements of double seats that could be made up in lower berths. Upper berths were concealed behind slanting panels in the upper walls. By pressing control buttons, seats and seat backs could be adjusted and made into berths in about the same time it took to prepare the upper berths.

Berths were equipped with call buttons, reading lights, and conveniences for the orderly storage of clothing and toilet accessories. In addition, the upper berth had its own air-conditioning controls. The lower was ventilated from the cabin. The upper also had handy compartments in the wall, one containing a waste disposal bin; below the bin was a small jewel case; the other had a thermos bottle, a large wax cup, shaving or make-up kit, and a small shelf with an adjustable mirror. At the foot of the berth was a large shelf for an overnight bag. Each upper berth had its own window.

DC-6 Hydraulic System

The hydraulic system provided the energy to operate the landing gear, the wing flaps, nose wheel steering, the wheel brakes, and the windshields wipers. A hydraulic

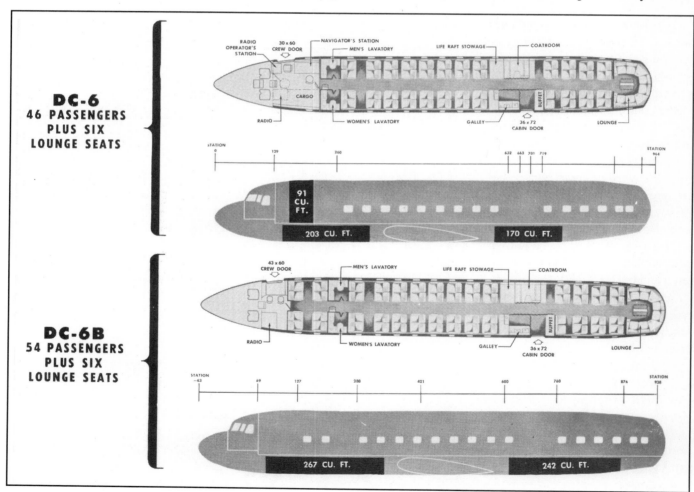

There were various seating arrangements available for the DC-6 and DC-6B. Shown is the standard dayplane overwater version. (Boeing Historical Archive)

Sabena's DC-6 aircraft were purchased to Type Specification 1159, which was an overwater day forward cabin and aft cabin sleeper. The use of thermal anti-icing system precluded the necessity of rubber boots attached to the wing and tail leading edges. (Boeing Historical Archive)

Although initially delivered without weather radar to American Airlines, this DC-6 of Aeronaves De Mexico has been retrofitted with the nose-mounted equipment. Also clearly shown is the dual high frequency communication antennas strung from the upper forward cabin to the empennage. (Harry Gann Collection)

reservoir containing 5.4 U.S. gallons of fluid supplied the 3,000-psi hydraulic oil pressure via pumps driven by the two inboard engines. There was also an auxiliary hydraulic pump for use during an emergency or during ground maintenance work.

In the event of failure of the hydraulic braking system, the emergency metered air system was capable of providing either full or partial brakes.

DC-6 Electrical System

The DC-6 electrical system consisted of a 24 to 28 volt, direct current, single-wire installation. The system provided power for starting, exterior and interior lighting, operation of cowl flaps, signaling and warning devices, and incorporated multiple busses for improved power distribution and protection against system failure. The structure of the aircraft formed the negative or ground return for the majority of the circuits.

Primary DC power was delivered to the electrical system from four engine-driven generators capable of an output of 300 amperes each at 28 volts. Two 12-volt, 88 ampere-hour batteries connected in series supplied auxiliary power for the 24 to 28 volt DC system. An external power receptacle was installed to permit ground power to be used.

Two rotary inverters located in the soundproof compartment supplied alternating current. Either inverter was capable of supplying the entire alternating-current load, but both could be connected to the system to divide the load, and supply 115-volt, 400 cycle AC current. Also 26-volt, 400-cycle AC current, both single and 3-phase, was furnished to the instruments through step-down transformers.

DC-6 Anti-Icing

A hot air anti-icing system, rather than rubber de-icer boots, was used to prevent the accumulation of ice on the leading edge of

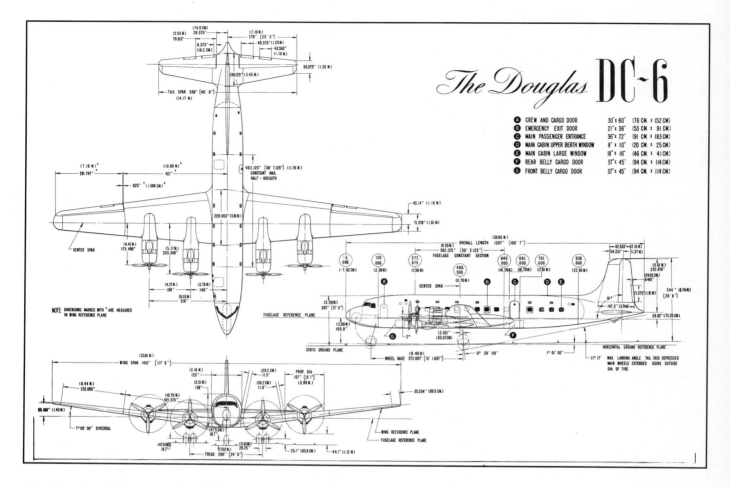

The Douglas DC-6 general arrangement shows that the passenger entry doors were placed exclusively on the left side of the aircraft. On the earlier DC-3 aircraft, the side of the aircraft where the doors were placed was optional. (Boeing Historical Archive)

the wings and empennage. Heated air was supplied through ducts to the upper and lower leading edge surfaces. This thermal system was completely enclosed in the airfoil surfaces, thus eliminating drag and lift variations experienced with de-icer boots. This was another first in commercial air transports.

Air entered the anti-icing system through a scoop located in the leading edge of the wing and the lower leading edge of the vertical stabilizer. A small part of the inflow was used as combustion air to mix with heater fuel. The remainder, heated to 400°F, flowed through ducts to the leading edges of the airfoil where it entered the narrow double skin, flowing chordwise through a series of small holes into the interior of the airfoil and then through the open airfoil section to a discharge outlet.

Three combustion-type heaters were used and were positioned one in each wing and one in the tail. They were interchangeable and had a maximum capacity output of 273,000 BTU per hour. The heaters were fired with gasoline supplied from the airplane's fuel system. The heaters and accessory containers were fire and explosion-proof, and were protected by the automatic fire extinguisher system. Any fuel fumes formed within the container were carried by combustion into the heater and burned with the fuel mixture. In case of a fire, CO_2 and the cutoff of fuel and ignition could extinguish the fire in the heater.

Twenty-five average sized houses could have been heated with the capacity of this heating equipment.

An electrical system was used to prevent ice forming on the propellers. The heater units consisted of rubber boots of a conducting material, which were cemented around the leading edge of each propeller blade. Electric power passed through brushes and rings to the propeller hub where the circuits were joined and connected through more brushes and slip rings to the engine case and thence to the main electrical bus. A timing device applied 3,000 watts to each set of blades in turn for about twelve seconds to dislodge any ice accumulation.

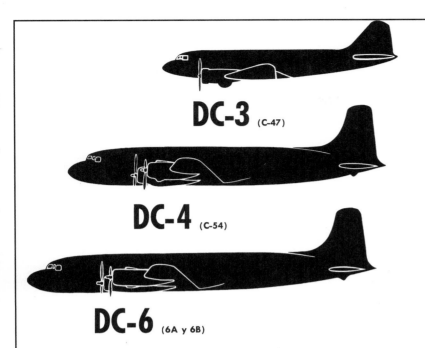

In 1954 Douglas placed this two-page advertisement in several Latin American periodicals. (Boeing Historical Archive)

Five DC-6 aircraft were delivered to Flota Aerea Mercante Argentina in 1948. All were built to the basic DS-477B specification. (Boeing Historical Archive)

The DC-6A had two large cargo loading doors in addition to the two lower fuselage cargo doors. The DC-6/DC-7 converted freighters also used this DC-6A design. (Boeing Historical Archive)

The first DC-6 for the Swedish airline SILA. It was an overwater, 52-passenger day, 26-passenger sleeper built to the basic DS-477 specification. The two wire antennas leading to the vertical stabilizer are high frequency communication antennas. Protruding from the nose is the glide slope antenna, while the wire antennas on the belly are ADF sense antennas. (Boeing Historical Archive)

DC-6 Versions

Air Freighters and Military Transports — Chapter 3

In addition to the basic DC-6, Douglas developed several variants to accommodate specific customer requirements.

DC-6 Military Versions

While the DC-6 is best known as a commercial airliner, it was initially proposed as a military transport aircraft, the YC-112A. Even though the U.S. government contributed to the cost of development and flight test, and purchased many production aircraft, the success of the series was based on the commercial applications. Also, many of the aircraft that were initially purchased by civil operators were later acquired for use by foreign military agencies.

The U.S. government ultimately purchased 167 DC-6-type transports and they were in the inventory into the 1990s. A few were shifted from the military to other government organizations for special missions. After separation from military usage, some are still in operation, where their particular economics make sense.

XC-112

This program was initiated in 1944 and was an extensive redesign of the C-54, including the use of the P & W R-2800C series engines. The development effort was cancelled later that year.

XC-112A

The XC-112A (later redesignated YC-112A as specified in DS-478) was a reactivation of the XC-112 with an 80-inch fuselage extension, and took the place of the XC-115. One YC-112A was built and delivered to the USAAF on June 22, 1948. Douglas used this airframe, with the government's concurrence, for some of the DC-6 certification testing. The airframe was then surplused to Conner Air Lines in 1956. It was destroyed in 1967 while being operated by Mercer Airlines when it made a forced landing after an engine failure in a golf course but caught fire from a spark created by a fireman's cutting tool.

XC-114

The XC-114 (DS-490) was to be powered by four liquid-cooled Allison V-1710 engines. The aircraft was to be equipped with a passenger interior. The program was cancelled in December 1945 before any aircraft were built. At that time, there was thought given to flight-testing the XC-112 with the Allison engines, but that plan was also abandoned.

American Airlines chose to eliminate the cabin windows in their ten Airfreighters. Some airlines left the windows installed to allow them the option of carrying passengers. (Boeing Historical Archive)

Slick Airways was the initial operator of the stretched, cargo-carrying version of the DC-6. With a total buy of 13 aircraft, Slick purchased more DC-6As from the factory than any other airline. (Boeing Historical Archive)

The introduction of the DC-6A freight-carrying aircraft also brought a new industry into being. Equipment to facilitate the loading and handling of cargo had to be developed and sold to the airlines. The upward swinging doors were operated by the emergency hydraulic system. (Boeing Historical Archive)

XC-115

The XC-115 was intended to be powered by liquid-cooled Rolls Royce (or Packard-licensed) Merlin V-1650 engines. It was to carry an ambulance-type interior. This program was cancelled in February 1945 in favor of the XC-112A.

XC-116

A Douglas proposal to build an XC-116 aircraft with advanced engines for the Army Air Forces was cancelled in October 1945 due to unavailability of the General Electric TG-100 turboprop engines that were supposed to power it.

VC-118 *Independence*

The *Independence* was a single DC-6 (C/N 42881) that was provided for the U.S. presidential fleet and used by President Truman. It was based on the American Air Lines configuration and early delivery was facilitated on July 1, 1947, when American gave up one of its delivery positions. In 1964, having being relegated to VIP usage after serving the president by the Air Force, the *Independence* became a display at the Air Force Museum at Wright Patterson AFB, on loan from the National Air and Space Museum.

C-118A

One hundred C-118A aircraft were purchased for use by the Military Air Transport Service (MATS), which later became the Military Airlift Command (MAC). These aircraft were generally similar to the DC-6A. Some were used as VIP transports or as aeromedical aircraft, but the bulk were used as airlifters. The first of these aircraft was delivered to the Air Force on July 16, 1952.

R6D-1/C-118B

The U.S. Navy was the first to purchase the military DC-6A when it ordered 65 aircraft, with the first delivery in 1951. When the Department of Defense altered the designation system for U.S. military aircraft in 1962, the R6Ds became C-118Bs. Forty of these aircraft were transferred to the Air Force when MATS was designated as the primary air transportation unit for all military services.

Other U.S. government agencies also used the DC-6, including NASA and the FAA, although they procured

The initial DC-6A Liftmaster was flown September 29, 1948, and used by Douglas for flight test, development, and marketing until it was delivered to Slick Airways in 1951. (Boeing Historical Archive)

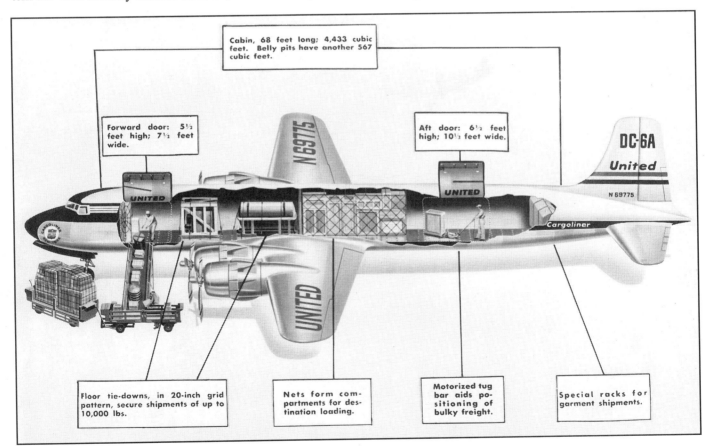

The United Air Lines DC-6A Cargoliner had two largo doors, one forward and one aft, to aid in the loading and positioning of mixed cargo. (United Air Lines via Harry Gann)

Overseas National Airways (ONA) operated three DC-6A aircraft before selling them to Aaxico in 1961. (Boeing Historical Archive)

DC-6A

To meet the increasing demand for an aircraft tailored to accommodate the hauling of non-passenger items, both for commercial and military operators, Douglas announced the development of "...a huge, new, modern air freighter" on July 1, 1948. It was to be a stretched version of the DC-6 called the DC-6A Liftmaster. (The Air Force designated their version the C-118A, while the Navy used the R6D-1 designation.) Actually, Douglas had started engineering the project on February 1, 1948, and authorized manufacturing on April 30, 1948.

The changes made to the basic DC-6 structure were relatively few to adapt it to the mission of carrying

DC-6 Airline Users

Prior to the maiden flight of the YC-112A, American Airlines became the first airline to order the DC-6 when it contracted for 50 aircraft. United Airlines followed shortly after with a firm order for 35 aircraft. Both airlines began full service with their DC-6s on April 27, 1947. Sabena, the Belgian airline was the first overseas operator. When the final DC-6 was rolled out and delivered to Braniff Airways on November 21, 1951, 175 aircraft had been delivered to commercial airlines.

these aircraft from other users. Many foreign governments also used DC-6s in military applications.

While delivered without passenger windows, this ex-American Airlines DC-6A was operated by World Airways as a convertible passenger-freighter. (Harry Gann)

ANA ordered four dayplane configured DC-6B aircraft for routes within Australia. In 1957 ANA was subsequently sold to Ansett to become Ansett-ANA (Boeing Historical Archive)

bulky goods. First, the fuselage was stretched 60 inches just forward of the wing. (This was an "easy" change as Douglas had designed the DC-4 and DC-6 with a constant section fuselage.) This increased the loadable volume and expanded the allowable center of gravity range at maximum payload, which permitted uniform loading over the entire cabin length. The available volume increased to 5,000 cubic feet, including the two lower cargo holds.

The DC-6A gross takeoff weight was increased from 97,200 to 107,000 pounds, the gross landing weight from 80,000 to 92,360 pounds, and the zero fuel weight from 74,000 to 87,360 pounds. A number of structural revisions to the fuselage were made to permit these increases The main cargo floor, which consisted of sheet metal panels, was rated at 200 p.s.f. static loading. Skin thickness revisions and some stringer beef-up, especially in the region of the large cargo doors were made. Also there was a general "beef-up" of the landing gear support structure.

The noticeable changes to the structure were the addition of large doors to permit the loading of large objects into the cabin area. The forward cargo door had an opening 67 inches high and 91 inches long, while rear door opening was 78 inches high and 124 inches long.

If the operator desired, the DC-6A could be quickly converted to a passenger-cargo combination to adapt to seasonal variations in traffic.

Original Airline Customers for DC-6 Aircraft

Airline	Quantity	Factory Serial Numbers
American Airlines	50	42854-42865
		42879-42880
		42882-42896
		43035-43054
		43137
Braniff Airways	9	43105-43110
		43293-43295
British Commonwealth Pacific Airline	4	43125-43128
Cia Mexicana De Aviacion	3	43211-43213
Delta Airlines	7	42897-42899
		43139-43140
		43142
		43219
Flota Aerea Mercante Argentina	5	43030-43033
		43136
Hughes Tool Company*	1	43152
Royal Dutch Airlines (KLM)	8	42900
		43111-43112
		43114-43118
Linee Aeree Italiane	3	43215-43217
National Airlines	8	43055-43058
		43150-43151
		43214
		43218
Pan American Grace Airways (Panagra)	6	42876-42878
		43103-43104
		43141
Philippine Air Lines	5	42902-42903
		43059-43060
		43138
Sabena	5	43062-43064
		43148-43149
Scandinavian Airlines System (SAS)	13	43119-43124
		43129-43135
United Airlines	48	42866-42875
		43000-43029
		43061
		43143-43147
* Sold to Linee Aeree Italiane prior to delivery		43148-43149
TOTAL	**175**	

To supplement a number of DC-6 aircraft, the Societe Anonyme Belge d'Exploitation de la Navigation Aerienne (Sabena) purchased nine DC-6Bs and one DC-6A. The DC-6Bs were initially operated with 77 reclining seats for tourist-class service, and were operated on Sabena's trans-Atlantic, middle East and African routes. (Boeing Historical Archive)

The floor tie-down fittings were arranged in a standard grid pattern so that seats could be installed. Some operators referred to aircraft used in this manner as DC-6C aircraft.

The first DC-6A was delivered to Slick Airways on April 16, 1951, which initiated operations with the aircraft five days after the introduction of the first DC-6B scheduled flights by United Air Lines. Other airlines soon followed with the introduction of all-cargo flights, including Pan American with flights over the North Atlantic.

The last DC-6A off the Santa Monica production line was fuselage number 1037 ordered by Loide Aereo Nacional (LAI) but delivered to Panair do Brasil December 12, 1958. A total of 74 of the cargo/com-

Original Airline Customers for DC-6A Aircraft

Airline	Quantity	Factory Serial Numbers
American Airlines	10	43839-43841
		44914-44917
		45373-45375
Air Liban	1	45226
Alaska Airlines	2	45243
		45517
Canadian Pacific Airlines	6	44063-44064
		45497-45500
Flying Tiger Airline	7	44071-44075
		44677-44678
Japan Air Lines (JAL)	2	44069-44070
Royal Dutch Airlines (KLM)	2	44076
		44257
Los Angeles Air Service	2	45480-45481
Nevada Aero Trades Co	2	45503-45504
Northwest Airlines	1	44890
Overseas National Airways	3	45474-45476
Pan American World Airways	3	44258-44260
Riddle Airlines *	1	45372
Sabena	1	44420
Slick Airways	13	43296-43297
		43817-43819
		44889
		45058
		45110
		45457-45458
		45518-45520
Swissair	1	45551
Trans Caribbean Airways	3	45227
		45368-45369
United Airlines	7	44905-44909
		45521-45522

* Originally sold to Olympic Airways, not taken up. Sold to Riddle but delivered to Hughes Tool Company where it was stored at Santa Monica for some 16 years.

TOTAL   67

One of two DC-6A aircraft flown by Alaska Airlines. Delivered in 1958, it was lost in an accident while approaching Shemya Island in 1961. (Boeing Historical Archive)

Some airlines, such as Loide Aereo, opted to call the convertible DC-6A aircraft DC-6C to reflect the diverse payload capability. (Boeing Historical Archive)

While originally ordered by Trans Caribbean, OD-ADC was actually delivered to Air Liban who chose to designate it a Super DC-6C. (Boeing Historical Archive)

Hunting Clan Air Transport Ltd, an English airline, ordered two DC-6C aircraft. The DC-6C had factory-installed equipment to expedite the time to convert between the payload options. (Boeing Historical Archive)

This United Air Lines' DC-6B was a standard dayplane version as specified in Detail Specification 1198. United ordered 43 DC-6B aircraft, receiving the last on June 10, 1958. (Boeing Historical Archive)

Pan American World Airways purchased more DC-6B aircraft than any other airline, ordering 45. Pan Am DC-6Bs were normally configured for 44 passengers, but had provisions to carry up to 85. (Boeing Historical Archive)

The Arabian American Oil Company (ARAMCO) purchased three DC-6B aircraft for business transportation. They were configured to Detail Specification 1225, which was overwater, forward cabin dayplane and aft cabin sleeper. Notice the three sleeper cabin windows in the aft fuselage. This aircraft, N708A, after being sold to Transair Sweden, crashed on September 18, 1961, killing UN Secretary General Dag Hammerskjold. (Boeing Historical Archive)

The Japan Air Lines DC-6B aircraft used Hamilton Standard reversible, hydromatic propellers, as did all of the DC-6Bs. The propellers were constructed from dural and were 13 feet 6 inches in diameter. JAL did not use the optional propeller spinners. (Boeing Historical Archive)

bination versions of the DC-6 were to be delivered to the world's airlines. The utility of the DC-6A as well as the surplus military version, the C-118/R6D aircraft, was greatly appreciated and many airlines were to ultimately use them.

DC-6B

To meet the continuing competition for passenger transports, mainly brought on by the Constellation, Douglas introduced the DC-6B. This proved to be a master stroke and this aircraft has been acknowledged as the finest piston-engined transport ever built. Its operating costs were the lowest of any reciprocating transport.

The DC-6B included the same 60-inch fuselage stretch as the DC-6A, providing seating for eight additional passengers in the deluxe configuration, and up to a total of 102 in coach versions. More luggage and cargo could be carried in the lower fuselage compartments. Deleted were the reinforced floor and the cargo doors. Takeoff power was increased to 10,000 horsepower. Although U.S. domestic airlines usually operated the DC-6B at a gross take-off weight of 100,00 pounds, these aircraft were structurally capable of operating at 107,000 pounds. Overseas operators utilized the additional weight to increase the fuel capacity, allowing one-stop trans-Atlantic flights with a full load of tourist passengers.

The DC-6B made its first flight on February 10, 1951. Many airlines that had not ordered the DC-6, quickly took advantage of the DC-6Bs capabilities and placed orders. In a repeat of the DC-6 scenario, American and United Airline took early delivery, with United placing the aircraft in service on April 11, 1951 followed by American Airlines on April 29, 1951.

Union Aeromaritime de Transport (UAT) ordered the optional propeller spinners on its two DC-6B aircraft, which were overwater, forward cabin dayplane and aft cabin sleeper configured. (Boeing Historical Archive)

The lower seat miles costs of the DC-6B proved to be an incentive for U.S. trunk carriers, and soon all but TWA ordered them. Pan American used them to inaugurate tourist-class service over the North Atlantic. Swissair became the first foreign DC-6B operator by placing them in service in July of 1951.

Production ended in 1958 after the delivery of 288 aircraft. This was also the last of the 704 DC-6/C-118/R6D series built. Jugoslovenski took delivery of this aircraft on November 17, 1958.

DC-6C

On June 19, 1953, Douglas announced a new addition to the DC-6 series that combined the special cargo capabilities of the DC-6A with the passenger features of the DC-6B airliner. The basic DC-6C interior could be changed in a matter of minutes from a configuration carrying 76 passengers and their baggage, plus 2,400 pounds of cargo, to an interior carrying nearly 13 tons of cargo.

A movable bulkhead, which separated cargo from passengers, was the chief key to the airplane's versatility. It could be positioned at four different stations, allowing an airline to alter the interior arrangement to suit varying passenger and cargo requirements on each flight. Large doors, fore and aft, provided maximum accessibility for loading cargo.

The DC-6C length of 105 feet 7 inches, wingspan of 117 feet 6 inches, and overall height of 28 feet 8 inches, was identical to the DC-6A.

Seven of these aircraft were delivered, although many DC-6A/Bs were later converted in the field to this configuration.

The DC-6C configured as a convertible passenger-cargo carrier was initially delivered to Hunting Clan Airways on August 7, 1958. A total

Original Airline Customers for DC-6B Aircraft

Airline	Quantity	Factory Serial Numbers
American Airlines	25	43263-43273
		43543-43547
		43564
		43845-43847
		44056-44060
Aigle Azur	1	44871
Alitalia	6	44251-44254
		44888
		44913
Australian National Airways	4	44693-44694
		45076-45077
Arabian American Oil Company (ARAMCO)	3	43559-43560
		45059
Continental Air Lines	3	44082-44083
		44689
Civil Air Transport	1	45550
Cia Mexicana De Aviacion	2	43836-43837
Canadian Pacific Airlines	12	43842-43844
		44062
		44891-44892
		45078-45079
		45326-45329
Cathay Pacific Airways	1	45496
Ethiopian Airlines	3	45523-45524
		45533
Japan Air Lines (JAL)	2	44432-44433
Jugoslovenski Aerotransport	2	45563-45564
Royal Dutch Airlines (KLM)	7	43550-43556
Linee Aeree Italiane	4	44417-44419
		45075
Linea Aerea National De Chile	7	44690-44692
		45513-45516
Maritime Central Airways	2	45505-45506
National Air Lines	9	43738-43743
		43820-43821
Northeast Airlines	10	45216-45225
Northwest Airlines	12	44698-44699
		45197-45202
		45319-45320
		45501-45502
Olympic Airlines	4	45539-45540
		45543-45544
Pan American World Airways	45	43518-43535
		43838
		44061
		44102-44121
		44424-44428
Pan American Grace Airways (Panagra)	4	43536-43537
		44255-44256
Philippine Air Lines	2	43557-43558
Sabena	9	43828-43832
		44175-44176
		44695
Scandinavian Airlines System (SAS)	14	43744-43749
		43548-43549
		44165-44170

Original Airline Customers for DC-6B Aircraft (continued)

Airline	Quantity	Factory Serial Numbers
Swissair	6	43274-43275
		43750
		44087-44089
Trans-American Airlines (North American Airlines)	7	44687-44688
		45107-45109
		45472-45473
Transports Aeriens Intercontinental	5	43833-43835
		44696-44697
United Air Lines	43	43257-43262
		43276
		43291-43292
		43298-43300
		43538-43542
		43561-43563
		44080-44081
		44893-44902
		45131-45137
		45491-45494
Union Aeromaritime de Transport	2	45478-45479
Western Air Lines	31	43822-43826
		44429-44431
		44434
		45060
		45063-45067
		45173-45179
		45321-45324
		45534-45538
TOTAL	**288**	

of seven aircraft were manufactured to this configuration. However, in the years after delivery, many DC-6A and DC-6B aircraft were converted to "QC" or quick-change configuration and were designated DC-6A/B rather than DC-6C aircraft.

DC-6A/B, DC-6A/C, DC-6BF, and DC-6B-ST

Many DC-6A or DC-6B aircraft were altered to convertible passenger-cargo configuration by Douglas-licensed Pacific Aeromotive Company as well as other engineering or airline shops. They were given various designations and the actual modifications varied with the ultimate users. Two such DC-6B's were converted to swing-tail (DC-6B-ST) by the Sabena Airline's engineering shop.

Shown in its house colors at a local airshow, this DC-6A reflected the marketing impact on the colors applied to the test vehicles. (Harry Gann)

This underside view of a Northeast DC-6B shows some of the numerous antennas. Forward centerline is a VHF communication antenna, with two ADF loop antennas flanking it. At the wing roots are ADF sense antennas. (Boeing Historical Archive)

Canadian Pacific Airways used 12 DC-6Bs on its routes from Vancouver to Australia and New Zealand via Honolulu and Fiji. Later, Canadian Pacific traveled over the North Pole to Amsterdam. (Boeing Historical Archive)

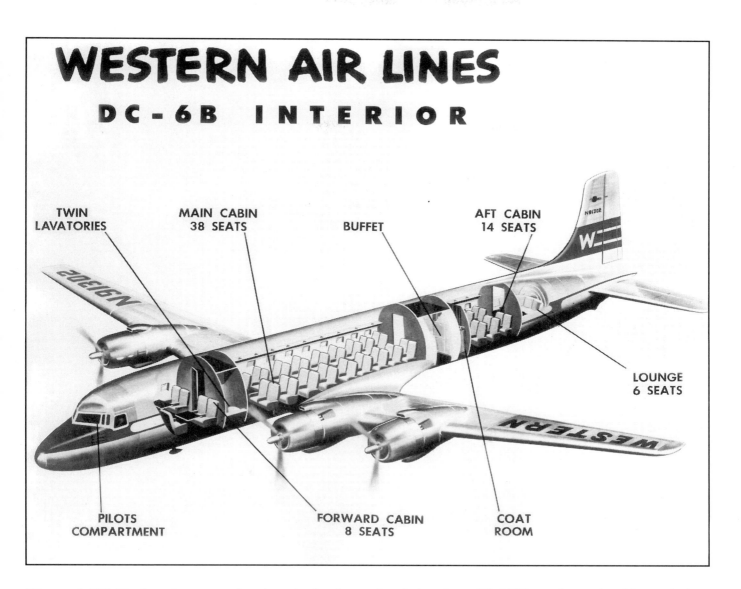

Western's DC-6B aircraft were ordered as the basic overland, dayplane DS-1198, as shown by this rendering. DS-1809 was the legal document that described the specific equipment particular to Western's requirements. (Boeing Historical Archive)

Most of the DC-6 and DC-7 series were delivered with lounges either forward or aft, or both. However, competition for lower seat mile costs forced the airlines to later substitute higher density seating. Shown is a DC-6B installation. (Boeing Historical Archive)

Cathay Pacific Airways was formed in Hong Kong in 1946, operating its only DC-6B generally in a southerly direction from Hong Kong. This particular aircraft eventually was to be modified to a fire fighter. (Boeing Historical Archive)

DC-6 aircraft were designed, manufactured, and assembled at the Douglas Aircraft Company main facility located at Clover Field in Santa Monica, California. Shown is the second of six Swiss Air Lines DC-6B aircraft delivered one month after the date of this photograph. Swiss Air was the first to operate DC-6B aircraft on the North Atlantic and the first to operate them outside the United States. (Boeing Historical Archive)

Pan American Grace Airways (Panagra) purchased four DC-6B aircraft of the 288 built by Douglas. The DC-6B featured special floor tracks running the full length of the cabin, which permitted quick relocation of the seats when a variation in seating density was desired. (Boeing Historical Archive)

Scandinavian Airline System (SAS), an amalgamation of airlines from Sweden, Denmark, and Norway, inaugurated the first Great Circle route from Copenhagen to Los Angeles via the North Pole in November of 1954 with DC-6B aircraft. Ten hours of travel time was saved over the route via New York City. (Boeing Historical Archive)

While SM5 was originally delivered to Alitalia in 1953, the DC-6B was acquired for military support by the Italian Air Force in 1970. (Tim Williams via Harry Gann)

DOUGLAS DC-6/7

National Air Lines acquired nine dayplane versions of the DC-6B for domestic routes. (Boeing Historical Archive)

Transports Aeriens Intercontinentaux was selected by France to be the sole French operator in the Pacific and Australia in 1956. The company procured the DC-6B for this route and used five aircraft. In 1963, TAI merged with Union Aeromaritime de Transport (UAT) to form Union de Transports Aeriens (UTA). This photograph shows a takeoff from Los Angeles International Airport on a nonstop, 5,700-mile flight to Paris. (Boeing Historical Archive)

North American Airlines changed its name to Trans American Airlines when the manufacturing company of the same name protested. NAA and TAA both traded as Twentieth Century Airlines and provided discount fares for domestic routes. (Boeing Historical Archive)

Linea Aerea National De Chile (Lan Chile) used its DC-6Bs to expand routes to Lima, Peru, and to Miami in the United States. Shown on a test flight over the Long Beach, California, harbor is the company's fifth DC-6B. (Boeing Historical Archive)

Trans American Airlines' markings were only slightly altered when the name was changed from North American Airlines. Their seven DC-6B aircraft were dayplanes built to DS-1198. (Boeing Historical Archive)

Olympic Airlines began operations in April of 1957 with a DC-4 but later purchased four DC-6B for its routes, covering Europe and the eastern Mediterranean area. The aircraft were configured to the overwater, forward cabin day and aft cabin sleeper versions. (Boeing Historical Archive)

Northeast Airlines operated Douglas DC-3 and DC-6B aircraft. Headquartered in Boston, Northeast used the DC-6Bs to expand service to New York, Montreal, Washington D.C., and to Florida cities. (Boeing Historical Archive)

N6521C was delivered to Pan American World Airways in 1952 but was leased to Capital in 1960 along with other PAA DC-6Bs. This dayplane, overwater version featured the optional propeller spinners as well as nose-located weather radar. (Boeing Historical Archive)

Ethiopian Airlines used three DC-6B aircraft in flights to Khartoum, Athens, and Frankfort. The DC-6B aircraft could carry up to 102 passengers in a cabin that was 8 feet 10 inches wide at the floor line. (Boeing Historical Archive)

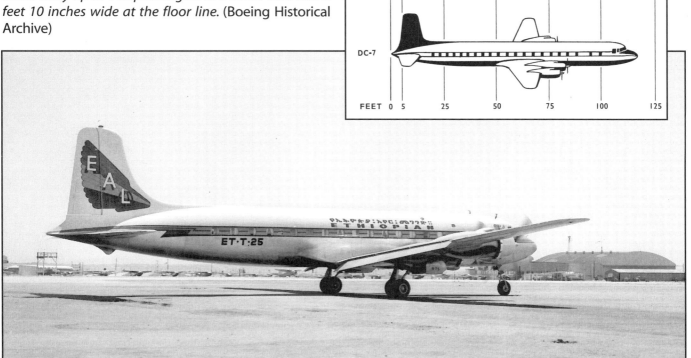

Australia National Airways DC-6B Beltana. ANA ordered four dayplane-configured DC-6B aircraft for their routes within Australia. In 1957 ANA was sold to Ansett to become Ansett-ANA. (Boeing Historical Archive)

The DC-6B, powered with the P&W radial, air-cooled engine that normally produced 2,500 hp, could takeoff at 107,000 pounds gross and cruise at 316 mph at an altitude of 23,000 feet. Northwest Airlines operated 12 of these aircraft on its over-the-pole and Pacific routes. (Boeing Historical Archive)

Original Airline Customers for DC-6C Aircraft

Airline	Quantity	Factory Serial Numbers
Hunting Clan Airlines	2	45531-45532
Loide Aereo Nacional	4	45527-45530
Sabena	1	44421
TOTAL	7	

The DC-7

Give the Customer What He Wants ... Chapter 4

In 1942 the Douglas plant at Santa Monica initiated design work on a large logistic transport for the Army Air Force designated C-74 Globemaster. It was described in design specification DS-425, and production was to be accomplished at the Long Beach facility. While the C-74 would have had a tremendous impact on the flow of supplies during the war, other Douglas aircraft such as the C-47 and the A-26, as well as the licensed Boeing B-17F/Gs, being built at Long Beach had a higher priority and the C-74 development suffered. The prototype did not fly until after the end of World War II, and only 14 were eventually built. The design was superceded by the very successful C-124 Globemaster II, which had a greatly enlarged fuselage.

After the war, the airlines were anxious to resume their commercial operations and to further expand to literally encircle the globe. On October 23, 1944, Pan American World Airways announced that it had conditionally contracted for a civil version of the C-74, to be designated the DC-7. Actually, Douglas had begun studies on this project when it issued DS-448 on May 11, 1942. The aircraft would carry 108 passengers with a crew of 13. Pan American would use the aircraft to greatly expand its service to Latin America, operating "clippers" directly from New York, Miami, New Orleans, and Los Angeles at fares as low as 3.5 cents per mile. To illustrate the size of these proposed aircraft, the news release stated that "their wings would be longer, from tip to tip, than a 16-story building and large enough for interior passage-ways for servicing and repairing engines in flight."

It was planned that these aircraft would have two spacious cabins, one

During the early stages of World War Two, Pan American World Airways contracted for a commercial version of the projected Douglas C-74 Globemaster logistics transport. Pan American envisioned that postwar transportation would continue the "clipper" style of luxurious accommodations which proved not to be the case. The design was abandoned and the designation "DC-7" was used on a later, different aircraft. (Boeing Historical Archive)

accommodating 72 passengers and the other 36, besides a galley equipped for serving full course meals, dressing rooms, toilets, cargo area, flight deck, and a rest compartment for the pilots. There would also be large storage compartments for food and electrically operated stoves and refrigerators.

The airline further stated that "The mammoth planes would carry passengers from New York to Buenos Aires in 22 hours, to Rio de Janeiro in 19 hours, and span the Pacific from Los Angeles and San Francisco to Honolulu in about 8 hours, making it possible for the everyday citizen to spend 12 days of his 2-weeks vacation abroad."

The Pan Am contract for 26 "four-engined clippers" at a cost of $40,000,000 was referred to as "the largest and most significant in the history of commercial aviation."

This proposed version of the C-74 Globemaster never made it into production. Pan Am's postwar thinking regarding transportation changed from the previous situation where luxurious and space-consuming items were available for the passenger's use. In view of the competition, this expensive clipper style was not realistic and Pan Am started its postwar expansion with the DC-4 and Constellation, later returning to the Douglas DC-6B and the DC-7s.

The DC-7

After the aborted Pan Am experience, the DC-7 designation was subsequently applied to a stretched version of the DC-6B. Douglas committed to building the DC-7 series in response to airline pressure and the competition from improved Constellation designs. The demand came first from the domestic airlines that wanted the capability of westbound nonstop flights and, in the case of the DC-7C, from the international airlines that desired increased over-water performance, both in range and speed. The performance of the Lockheed Constellations also impacted Douglas management decisions. The successful marketing effort of the DC-6B had helped immensely in bringing a reasonable profit to the stockholders. For the DC-7 series to be successful, the Wright R-3350 with the turbo compound exhaust system had to be commercially viable, providing both reduced fuel consumption and increased power. Complicating the production decision was the possible introduction of turboprop propulsion into the design of a new generation of transport aircraft. Douglas was somewhat reluctant to take on a new project, especially one that was based on an innovative and complicated addition to an engine that had not proven as reliable as the R-2800.

Primarily at the urging and financing of Cyrus "C.R." Smith of

As it was in many other instances, American Airlines was the pacing airline in the development of the DC-7, which was a stretched DC-6B with Wright R-3350 Turbo-Compound engines replacing the P&W R-2800 Double Wasp engines. American put the DC-7 in service on November 29, 1953, in direct competition with TWA's Constellations. (Harry Gann)

Initially, United Airlines did not have the confidence in the DC-7 operations that American Airlines showed. However, UAL ordered 25 DC-7s in 1952 and put them in service on domestic routes on June 1, 1954. United eventually ordered 57 DC-7s. (Boeing Historical Archive)

American Airlines, Douglas began development of the DC-7 in the spring of 1951. American contracted for the aircraft in December 1951. The aircraft made its first flight on May 18, 1953, and was certificated by the CAA on November 12, 1953. The DC-7 was placed in commercial service by American Airlines in the first nonstop operations between Los Angeles and New York, in both directions, on November 29, 1953.

The DC-7B was the fastest of the Douglas four-engine transports that were powered with reciprocating engines. One was clocked at 410 mph at the Salton Sea, California, speed course. But in terms of costs per seat-mile, the DC-7 was more expensive to operate than the DC-6B. The Wright R-3350 engine never approached the reliability of the P&W R-2800 engine used in the DC-6 series. This also limited the DC-7 resale value. A few were converted to freighters, but most were relegated to serving as spare parts after the major airline users switched to turbine-powered aircraft. A few became air tankers because of the low initial acquisition cost on the used plane market. The U.S. military did not purchase any of the DC-7 series.

A direct descendant of the DC-6B, the DC-7 made history by making possible west-bound nonstop flights across the United States. Surprisingly, the DC-7 was profitable for Douglas from the delivery of the first airplane. It was generally regarded as an economically viable airplane despite the fact that the Wright engines were never entirely satisfactory.

DC-7 Design

Titanium, a heat-resistant and lightweight metal, was introduced to commercial transport design by Douglas on the DC-7. Areas where this "exotic" material were used included the structural covering of the aft nacelles, some frames, and the landing gear doors. Unalloyed titanium was also substituted for the stainless steel formerly used for firewall webs. The principal advantage of the new metal was that it had the strength of steel yet weighed only 56% as much. Use of this metal resulted in weight savings of approximately 200 pounds per air-

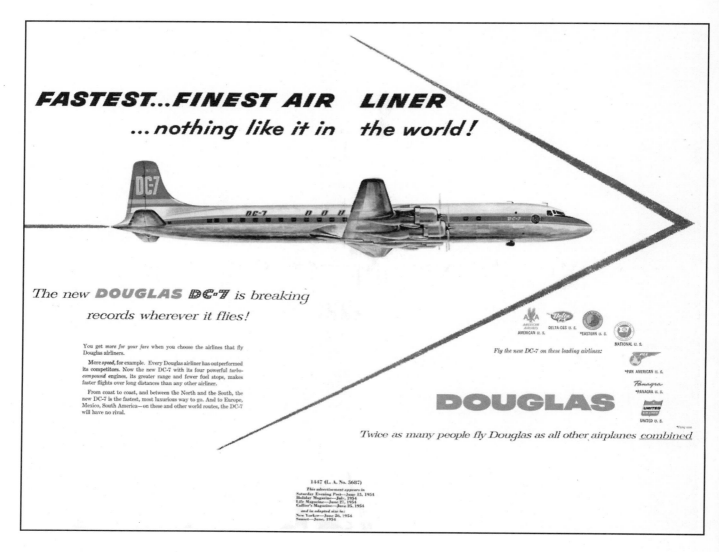

During the 1950s, Douglas was the top producer of commercial transports, and let the world know through a series of advertisements similar to this one. (Boeing Historical Archive)

plane, equal to one passenger and his luggage.

All control surfaces of the DC-7 were powered by flying tabs. Since power was always available from the air stream, no standby or auxiliary power boost arrangement was required for this system. As in the DC-6, the rudder was fabric covered, while all other control surfaces were covered with aluminum alloy.

DC-7 Cabin

The DC-7 and DC-7B fuselage length was increased by 40 inches over the DC-6B by adding a plug just aft of the trailing edge of the wing. Other than the insertion of the fuselage plug, the fuselage retained all of the major features and general contours of the DC-6 series, including the windows, passenger and crew doors, belly compartment doors, cockpit, and most on the interior items. Some operators utilized the additional cabin volume for cargo only, while others preferred to take advantage of the lower seat-mile costs by adding an additional row of passenger seats.

Despite the fuselage plug, it was determined that the DC-6B tail was sufficient to balance the airplane from a center of gravity standpoint, as well as provide the aerodynamic tail power required by the higher operating weights, speeds, and engine power.

The DC-7C fuselage length was stretched over the basic DC-7 an additional 40 inches ahead of the wing to give a length of 112 feet 3 inches. The main section of the fuselage extended from station 29 to station 978, which included part of the flight compartment, the forward compartment, twin forward and aft lavatory compartments, passenger cabin, and buffet on the main deck; and the hydraulic accessories compartment, air conditioning equipment compartment, and the forward and aft lower cargo compartments below the floor.

Passenger layout varied with the operator but a typical configuration provided two-abreast seating for 62 passengers in three major sections: eight in the forward compartment, 38 in the main cabin, and 16 in the aft cabin. For sleeper service, an alternate arrangement of the rear compartment accommodated 12 passengers in four upper and four lower berths. High-density interiors varied in capacity from 80 to 99 passengers. Below the floor cargo capacity was 651 cubic feet.

A special effort was made to isolate the sound-producing equipment such as engine exhaust, propellers,

To ensure the passengers comfort, 1,200 pounds of sound deadening material was installed in the DC-7 series. (Boeing Historical Archive)

The DC-7 cabin length was increased 40 inches over the DC-6B. Here, DC-7 fuselages clearly show the pressure dome located at station 978. (Boeing Historical Archive)

heaters, turbines, hydraulic accessories, etc. In addition, there was 500 pounds of sound deadening material used over and above the 700 pounds specified for the later DC-6B aircraft. Adding a third pane of glass, free-floating in rubber, to reduce noise, changed the passenger widow's design.

DC-7 Cockpit

The instrument panel and control layout was almost identical to the DC-6B, which allowed cockpit standardization for airlines that used both versions. Additions included a Bendix RMI, a primary direction indicator, and a Mach airspeed indicator which provided a continuous indication of actual, indicated, airspeed and the maximum airspeed at which the airplane should be flown.

DC-7 Wings

The DC-7 wing design was similar to that of the DC-6B except for some structural strengthening to accommodate the higher gross weights and performance levels. Both the inboard and outboard engine nacelles, however, were of a new design. The large increase in the thrust and torque of the new engines, plus the increase in the weight of the entire power plant unit, necessitated new engine section designs. An additional trim tab was provided in the aileron system for finer lateral trim.

For the DC-7C, 10 feet was added to the wing center section, making the overall wing span 127 feet 6 inches. Gained with this increase was a greater aspect ratio (9.93) and wing area (1,637 sq. ft.), plus additional space for fuel. Also, since the inboard engines were moved outboard, the passenger noise level was reduced.

DC-7 Empennage

Other than a longer rudder tab and a revision to the rudder control system to provide lower pilot forces, the DC-7 tail was similar to the DC-6B. On the DC-7C, the height of the vertical tail was increased by about two feet.

DC-7 Engines

A major change was made in the DC-7 power plant installation. The P&W R-2800, one of the most reliable piston engines ever produced, had reached its practical limit for increasing power output. The basic Wright R-3350 compound engine selected for the DC-7 could deliver 3,250 horsepower, an increase of approximately 20% over the latest DC-6B, with no increase in fuel consumption. This radial, air-cooled, 18-cylinder engine was equipped with a single-stage, two-speed, integral supercharger. A unique system of exhaust energy recovery provided the principal advantage of this engine by developing a higher power per unit of fuel burned. Each engine directed its exhaust gases through three integral turbine hoods, and a velocity turbine mounted in each hood returned power back into the engine crankshaft through a gear train and a fluid coupling. Heavy steel plating encompassed the turbine hoods to act as armor in case of failure of the turbine blades.

A new design feature was the introduction of an innovative airscoop to prevent ice-inducing moisture particles from entering the carburetor. This alternate air intake was used during icing conditions. The conventional airscoop, which had the stan-

An anomaly. DC-7 aircraft were not used in the U.S. military forces but DC-6A/Bs were under the C-118/R6D designation. This former United Air Lines DC-7, N6535C, was painted to represent a MATS C-118A in 1964 for a motion picture. (Harry Gann Collection)

The DC-7 was powered with the Wright R-3350 Turbo-Compound engine and driven with a Hamilton Standard constant speed, governor controlled, hydraulically positioned propeller. (Boeing Historical Archive)

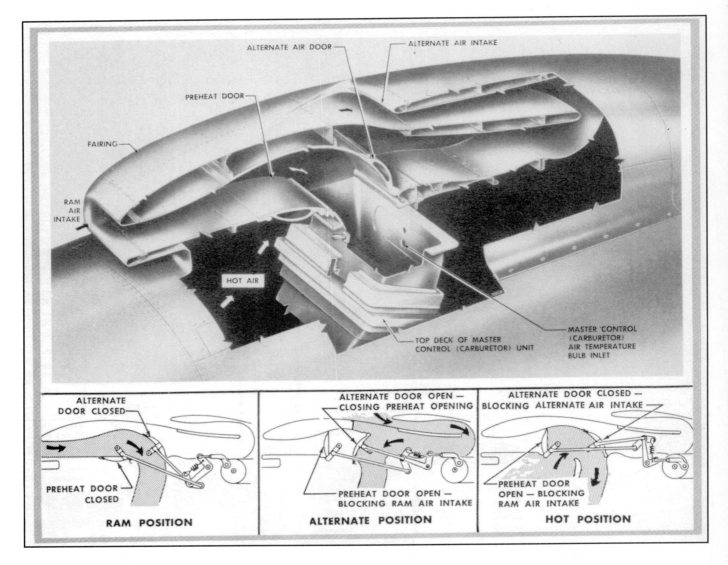

The DC-7 engines could draw induction air from three different sources: the standard ram scoop, an alternate ram scoop, or through the engine section as pre-heat. (Boeing Historical Archive)

dard inlet and hot air inlet, was used during normal operations.

The DC-7 and the DC-7B were equipped with Wright 972TC18DA1, 970TC18DA2, 972C18DA4, or 988TC18EA1 engines, all commercial versions of the R-3350 engine. The 18DA-series produced 3,250 brake horsepower at takeoff for 2 minutes, while the EA1 was rated at 3,700 bhp.

Either Wright 988TC18EA1or EA4 engines were installed on the DC-7C. The 18EA1 was rated at 3,700 bhp for takeoff using water injection, while the EA4 was rated at 3,400 bhp at takeoff dry.

BDC-7 Air Conditioning System

The pressurization system installed on the DC-7 was similar to that on the DC-6B and was supplied by two engine-driven superchargers. One internal combustion heater supplied cabin heat in-flight or on the ground. The DC-7 increased the airflow from 100 pounds of air per minute to 112, producing a complete air change in the cabin every two minutes. A five-ton capacity Freon refrigeration system was provided for both ground and in-flight. Adding 500 pounds to the aircraft weight, this refrigeration system eliminated the need for ground air conditioners.

DC-7 Anti-icing System

The anti-icing system for the airfoil surfaces was basically identical to that installed in the DC-6A and DC-6B aircraft. Propeller anti-icing was different only to the extent that the propellers were four-bladed rather than three. Windshield anti-icing did not employ alcohol; in fact some DC-6B airplanes were also delivered without the provisions of alcohol use.

DC-7 Electrical System

There were two main changes in the electrical system. The generator system incorporated 400-ampere General Electric generators rather than 350 ampere units, and the ground power receptacle supply was changed to provide a 1,000-ampere maximum capacity.

DC-7 Fuel System

The DC-7 and DC-7B used an eight-tank main arrangement similar to the DC-6B. However the total capacities could vary from 4,512 U.S. gallons to as high as 6,378 U.S. gallons. Included in the possible DC-7B tankage were external saddle tanks located on the upper wing surfaces as part of enlarged engine nacelles, as well as an optional 90-gallon tank situated in the center wing trailing edge. This DC-7B option could total 978 gallons of fuel. The DC-7C also used an eight-tank system, however 7,824 U.S. gallons of fuel could be carried.

Retractable dump chutes were installed (one in each nacelle) for the purposes of jettisoning fuel when necessary. The dump chutes and dump valves for all main and alternate tanks were controlled through cable rigging by the four levers located beneath a hinged floor plate aft of the control pedestal. Each lever had three positions: CLOSED (chutes up, dump valves closed), DRAIN (chutes partially extended, dump valves closed), and OPEN (chutes fully extended, dump valves open). Standpipes in the main tanks prevented the dumping of all fuel. An approximate 45-minute supply of fuel at 75-percent METO power remained in the tanks after dumping.

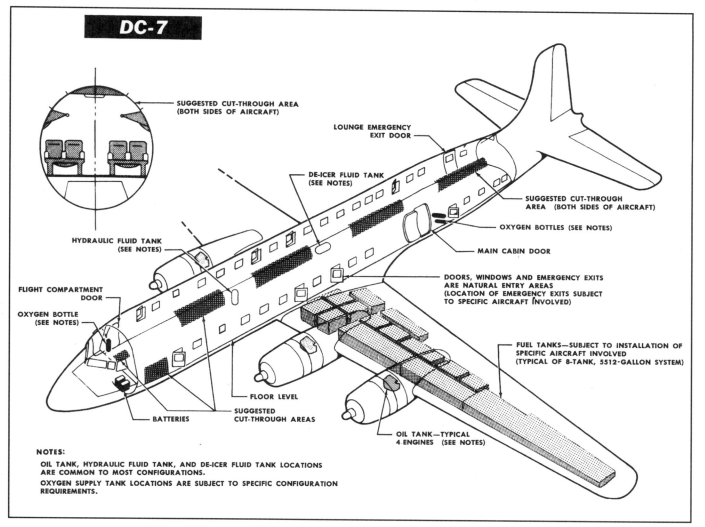

While this chart shows the DC-7 fuselage emergency access areas, it also shows the emergency doors and the layout of the fuel tanks. All fuel in the DC-6 and DC-7 aircraft was carried in the wings, although the exact configuration of the tanks differed among the airlines. (Boeing Historical Archive)

DC-7 Oil System

The DC-7 nacelle oil tanks were made large enough to hold all the oil required by the use of the wing carried fuel. This permitted the deletion of the DC-6B-type of oil transfer system, which weighed approximately 140 pounds and held 176 pounds of oil. The new nacelle oil tanks had a usable capacity of 160, 184, or 226 gallons, depending on the options selected by the customer. increased by a 10-foot addition between the wing roots, the Hamilton-Standard propeller diameter was increased to 14 feet on that version.

The propeller feathering system was actuated by an engine oil system that was configured such that the feathering system had priority on the use of the oil, thus allowing the pilot to feather the propellers even after a loss of engine oil. On some aircraft, an automatic propeller feathering system was installed that permitted

DC-7 Landing Gear

The landing gear group was composed of two fully retractable main gear units with dual wheels, and a fully retractable nose gear with a single steerable wheel. All were extended and retracted by hydraulic actuating struts, which were controlled by the landing gear control lever in the flight compartment. In the event of hydraulic failure, the gear could be lowered by gravity by

Overseas National Airways operated the gamut of Douglas piston engine aircraft. Shown is a DC-7. Overseas also operated DC-6A, DC-6B, DC-7B, and DC-7C aircraft. (Harry Gann)

DC-7 Propellers

To absorb the high amount of power that the R-3350 engines produced, specially contoured Hamilton-Standard four-bladed propellers were installed. The diameter was limited to 13.5 feet on the DC-7 and DC-7B by the proximity of the engine to the fuselage, the same as the DC-6B. To increase the diameter of the propellers would have also raised the sound level in the passenger cabin. Since the DC-7C wing span was

automatic feathering of a single propeller during take-off in the event of an engine failure.

Aerodynamic propeller spinners which improved the engine cooling system and provided less drag were available for installation on the DC-7B and DC-7C aircraft. By allowing a decrease in cowl flap angle, the spinners also further decreased drag, making possible higher airspeeds at given power settings. The spinner assembly was statically and dynamically balanced and serialized as a unit.

placing the landing gear control lever in the DOWN position to unlatch the uplocks. This allowed the gear to extend by its own weight, aided by the slipstream.

The main landing gear design on the DC-7 series was innovative in that it could be used as a speed brake to slow the aircraft during steep descents. Because of the clean aerodynamic design necessary for high speed cruising, it would have been impossible to make rapid descents from high altitudes, thus forcing the

After initially seeing service with United Airlines, N6314C was passed on to various non-scheduled airlines. ASA International operated it in early 1963. Later that year it was flown by Standard Airways. (Harry Gann Collection)

pilot to start his landing approach far from his destination. This could have caused some passenger discomfort by flying through the turbulence frequently found at lower altitudes.

To solve this vexing problem, the DC-7 main landing gear could be used as a speed brake and could be lowered at airspeeds up to 300 mph indicated airspeed (IAS), or up to 410 mph at 20,000 feet. The gear and related components such as struts, links, attachments, and landing gear doors were strengthened to permit this drag doubling force.

A faired, nonretracting tailskid, supported by a shock strut, protected the fuselage tail section from possible damage in the event of a tail-down landing or over-rotating during takeoff.

DC-7 Wing Flaps

On the DC-7B and DC-7C, an improved flap linkage system which gave the airplane higher lift characteristics for takeoff at greater payload was incorporated. The operating mechanism differs from the earlier versions primarily in that it placed the flaps into a more efficient position to provide optimum lift. At the same time sliding covers closed the gaps in the flap vanes at the three hinge stations on each wing to reduce drag. These were combined with vertical plates between the vane and flap nose and with hinge fairings on the bottom of the wing to reduce turbulence. The maximum flap deflection was 50 degrees.

DC-7 Crew Accommodations

The normal crew for a DC-7 series was a pilot in charge, co-pilot, flight engineer, a radio operator and/or a navigator, and two flight attendants.

DC-7C Weather Radar

Weather radar was offered on the DC-7C aircraft to facilitate greater schedule stability and increased aircraft utilization, and for raising the passenger comfort level. It was an outgrowth of a military navigational aid that had been refined and redesigned to show the pilot the weather ahead of the aircraft. Weather radar was also retro-

Delta Air Lines established DC-7 service on April 1, 1954, after merging with Chicago & Southern Air Lines May 1, 1953. These DC-7 aircraft were configured to carry 69 passengers, five in the rear lounge, 14 in the aft cabin, 42 in the main cabin, and eight in the forward "Sky Room." The operated on daily flights between Chicago and Miami. (Boeing Historical Archive)

fitted to some earlier DC-6 and DC-7 aircraft.

Weather mapping used radar equipment to project an electronic signal outward from the aircraft nose-mounted antenna to distances of up to 150 miles. The signal was reflected from the storm center ahead onto a monitor on the instrument panel. The pilot thus forewarned could navigate a course of least turbulence to the aircraft and with a minimum of time and fuel-consuming detours.

At that time, weather radar fell into two general classes: X-band (3.2-cm wavelength) and C-band (5.7-cm wavelength). The X-band was considered superior for ground mapping, beacon navigation, and weather mapping, but greater storm penetration and definition was provided by the C-band. Again, the option was left to the particular airline user.

The DC-7 series had strengthened main landing gear which acted as a speed brake to facilitate rapid descent from flight altitude into landing area. The nose gear would remain up during this maneuver. (Boeing Historical Archive)

The DC-6/7 in Color

All Over The World

The pre-war Douglas transports were, for the most part, finished in natural aluminum with only the user airline identification marks carried. There was good reason for this lack of color: weight savings. More payload could be carried if the application of paint was kept to a minimum. However, natural aluminum offered little protection from skin corrosion and that could eventually become a maintenance problem.

In postwar aviation transportation, where marketing effort plays a major part in the companies' operation, the colors used on the aircraft help reflect the image of the particular airlines. The more established and conservative airlines were reluctant to adopt the use of garish colors and markings. As the older officials retired and newer personnel came in, this attitude softened. Markings were changed from traditional to a conspicuous scheme to emphasize a redirection in the airlines' promotional effort. In some instances, well-known artists were commissioned to design an eye-catching arrangement.

Starting with the DC-6A, Douglas Aircraft also began to paint its prototype aircraft with distinctive colors and markings to promote the Douglas products. This was also a factor in the print advertising where illustrations of Douglas aircraft were used to sell the public on the benefits of aviation transportation and the virtues of Douglas-built aircraft.

Two DC-6A aircraft were modified to perform research into the actions of hurricanes. The Weather Bureau would penetrate these storms and record various parameters in an effort to predict the path and the effect of the hurricane. (NOAA via Carla Wallace)

In a joint delivery ceremony, American and United Air Lines took initial delivery of their first DC-6 aircraft on November 24, 1946. After pilot training and route proving, both airlines entered service with these aircraft on April 27, 1947. American inaugurated its service on the New York-to-Chicago route while United started with one-stop, 10-hour, transcontinental flights. (Boeing Historical Archive)

Conair Aviation Ltd, located at Abbotsford, Canada, has converted DC-6B aircraft into various non-airline uses such as fighting insect infestation, controlling oil spills, and fighting fires. Shown is the first DC-6B (CF-PWF) which could drop 3,000 U.S. gallons of fire retardant from an externally mounted tank. All versions of the DC-6 and DC-7 were used in this role. (Harry Gann)

Super Snoopy, the Racer. This ex-American DC-7B flew in a closed-course race with World War II "warbirds," finishing 6th out of a field of 20 aircraft. (Harry Gann)

Douglas used a former UAL DC-7 aircraft to run landing gear test for a possible application to the Air Force C-5 logistics transport proposal. (Boeing Historical Archive)

For the DC-7 rollout ceremonies, Douglas gathered together, from the top, a DC-3S, a DC-4, and a DC-6B for a photographic session with the first DC-7 (bottom). (Boeing Historical Archive)

National Airlines operated both the DC-7 and the DC-7B aircraft to supplement its DC-6 and DC-6B for a total of 25 aircraft bought from Douglas Aircraft Company. This DC-7B was the first of that type for National. (Harry Gann Collection)

T&G Aviation used two DC-7C aircraft in a contract to spray for locust infestation in Dakar, Senegal, Africa. Rebel forces shot down one of the two, hitting both with shoulder-mounted missiles. Although hit in one of the engines, this DC-7C was able to safely land. (Tim Williams via Harry Gann)

Douglas developed the DC-6A, initially as a dedicated freighter. However many airlines chose to operated these aircraft as a convertible passenger/cargo carrier. After being replaced by turbine-powered aircraft, many DC-6B aircraft were also modified to either freighters or combination freighter/passenger aircraft. (Harry Gann Collection)

DC-7 Versions

... But Don't Lose Money
Chapter 5

The prime consideration in the development of the DC-7 from the proven DC-6 series of transports was the means of obtaining added speed, range, and operating economy with better comfort and safety for the passengers. Nearly 500,000 engineering man-hours were spent in this effort.

Bearing a strong external resemblance to the DC-6 series of transports, the DC-7 was more than eight feet longer than the DC-6 and seated between 54 and 99 passengers, depending on the seating arrangement selected by the airline. The DC-7 had a top speed of more than 400 mph, with a normal cruising speed of 365 mph, 50 mph faster than the DC-6B.

Certification testing was carried out over six months using three aircraft. They made 204 flights in every type and condition of weather, flying the equivalent of nearly eight times around the world. Final approval came in 22 months after the initial design drawings were released to production.

DC-7A

Douglas bypassed the DC-7A designation during the initial production because it was felt that confusion would be generated. On the DC-6 series, the "A" was applied to the freighter version and when an improved version of the DC-7 was offered to the airline customers, it was called the DC-7B. Later, United Airlines renamed their DC-7s as DC-7A when they were converted to fly cargo payloads, a designation in the spirit of the DC-6A.

The DC-7B was a longer range DC-7, with higher gross weight allowable than the basic DC-7, that was defined in Detail Specification 1289. There were some variations in the configurations of the 112 aircraft delivered as DC-7Bs, of which this is the first aircraft. (Boeing Historical Archive)

DC-7B

The airlines' ever-increasing range requirement led Douglas to engineer changes to the basic airframe to accommodate this customer demand. The longer-range version of the DC-7 was called the DC-7B and was defined as an airplane whose operation weights were higher than those originally licensed on the DC-7. The old weights varied from 114,600 to 122,000 pounds maximum takeoff, and 95,000 or 97,000 pounds landing weight.

The specific changes required for the DC-7B nomenclature included DA4 or EA1 engines, improved wing flaps, and the higher structural weights. Installation of saddle tanks was optional on the DC-7B, as was the use of propeller spinners. The maximum gross take-off weight of over-water or intercontinental versions varied between 122,200 and 126,000 pounds.

South Africa Airways was the only non-U.S. airline to order DC-7Bs, operating four aircraft to London. (Boeing Historical Archive)

Pan American World Airways kicked off the DC-7B orders with seven aircraft purchased. They put this longer range DC-7 aircraft in service on the non-stop, New York-to-London route on June 13, 1955. Pan American's DC-7B had the full extended saddle fuel tanks on the engine nacelles, as did the South African Airways aircraft. (Boeing Historical Archive)

When Continental Airlines was awarded the rights to fly from Los Angeles to Chicago by the CAB, the airline ordered six DC-7B aircraft. This action moved Continental into the ranks of a major U.S. trunk airline. Notice the abbreviated engine nacelle saddle tanks that Continental chose for its DC-7Bs. (Boeing Historical Archive)

Shown with dual nation registration markings, this ex-Continental DC-7B was sold to Aerovias Ecuatorianas (AREA) in 1964. (Harry Gann Collection)

Pan American Grace Airways operated mainly in South America but did, through a three-way interchange with National and Pan American, provide service to New York City in its DC-7B aircraft. (Boeing Historical Archive)

Four DC-7B aircraft joined National Airlines' four standard DC-7s in operating between Florida and New York. National was also a Constellation operator. (Boeing Historical Archive)

The Swedish Red Cross operated this former American Airline's DC-7B in support of humanitarian efforts. (Tom Williams via Harry Gann)

Eastern Air Lines, a postwar Constellation operator, returned to Douglas when it placed orders totaling 50 DC-7B aircraft. Since Eastern did not need the maximum range version, the aircraft had short nacelle fuel tanks. The first 30 EAL DC-7Bs were equipped with cabin seating for 64 passengers, while the last 20 were fitted with 93 seats for coach service. (Boeing Historical Archive)

The Douglas DC-7B was publicly announced in December of 1953 when Pan American placed an order for seven aircraft. The DC-7B first flew April 21, 1955 and received its type certificate on May 27, 1955. It was placed in service by Pan Am on June 13, 1955, on a non-stop flight from New York to London.

Many of the DC-7Bs were reworked to haul freight after their service carrying passengers with the major airlines. Others were picked up by charter travel clubs and some were converted to air tankers to fight fires.

The Chicago White Sox baseball team used this ex-Eastern Air Lines DC-7B to meet its schedule commitments in 1964. (Harry Gann Collection)

Delta Air Lines ordered 11 DC-7B aircraft for 1957/58 delivery. The seating of these aircraft varied from 69 first-class arrangement to 90-passenger coach class. They were the last piston engine transports ordered by that airline, finally being retired in 1968 (Harry Gann Collection)

DC-7BF

This was a DC-7B converted to freighter configuration. See also DC-7F

DC-7C

When the Wright R-3350-series E engine became available, Douglas took the opportunity to investigate a further stretch of the existing airframe to make room for additional fuel. The additional power of the E series engine allowed a design that could fly the North Atlantic nonstop

In a superb marketing ploy, the DC-7C was named "the Seven Seas." The DC-7C was the ultimate development of the Douglas four-engine transports and it carried the bulk of the North Atlantic and polar traffic until the introduction of the Boeing 707 in 1959. (Boeing Historical Archive)

with full passenger/cargo loading under the most adverse winds. It has been said that the Douglas engineers visited the production area and concluded that the wings could be lengthened up to ten feet and still fit in the assembly area. A 10-foot plug was therefore installed in the wing center section, making room for almost 1,000 gallons of additional fuel. This permitted a range of 5,100 miles under ideal conditions. The wing stretch also moved the engines outboard, thus reducing noise, another plus for the configuration.

The DC-7C was one of the aerodynamically cleanest transports ever built. The parasitic drag was no greater than that of a cylinder 2.2-inches in diameter, the same length as the wingspan. This low-parasite drag was attained by various design refinements, including the installation of engine cowl liners and aerodynamic spinners, which encased the propeller hubs, reducing drag while making more efficient use of engine cooling air. Two ADF loop antennas that were formerly located externally were enclosed with semi-flush fiberglass fairings and the rear navigational light was streamlined to reduce the drag to 1/5th of the previous installation.

Market conditions permitted the introduction of the DC-7C and continued the production line until December 10, 1958, when the last of 1,042 DC-6/DC-7 series aircraft, a DC-7C, was delivered to KLM.

After making its first flight December 20, 1955, the "Seven Seas" received its airworthiness certificate on May 15, 1956, and entered service with Pan American World Airways on June 1, 1956. The DC-7C proved to be the most cost efficient of the aircraft used on the transoceanic routes and carried the

Pan American World Airways bought more DC-7C aircraft than any other airline. Shown is the first aircraft, which remained at the factory for ten months to participate in the flight test program. Flight test instrumentation can be seen mounted on each wing tip. The flight and certification testing required 200 hours in the air, the equivalent of more than 60,000 miles of monitored flight with three different aircraft. (Boeing Historical Archive)

After the completion of its service as a passenger transport, T & G Aviation who was contracted to spray for locust infestation in Africa eventually operated N284. While on a ferry flight, N284 was shot down by a rebel firing a shoulder-mounted missile, killing the five-crew members. (Harry Gann Collection)

The Royal Dutch Airline (KLM) operated 1049C and 1049G Constellations as well as DC-7C aircraft on its Polar and North Atlantic routes. KLM had inaugurated scheduled North Atlantic flights in 1946 using unpressurized Douglas DC-4 aircraft. Note the "Flying Dutchman" markings on the fuselage. (Harry Gann Collection)

The R-3350 power recovery system used turbines geared to the crankshaft and utilized the velocity energy of the exhaust gas. The power recovery system enabled the turbo-compound engine to achieve higher power with lower specific fuel consumption. (Boeing Historical Archive)

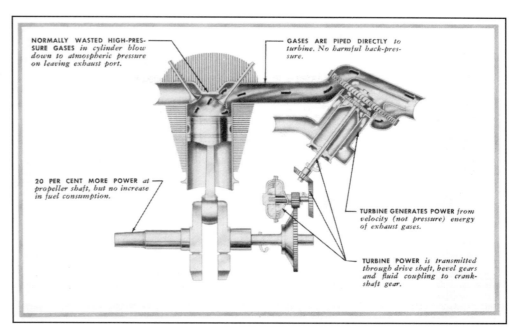

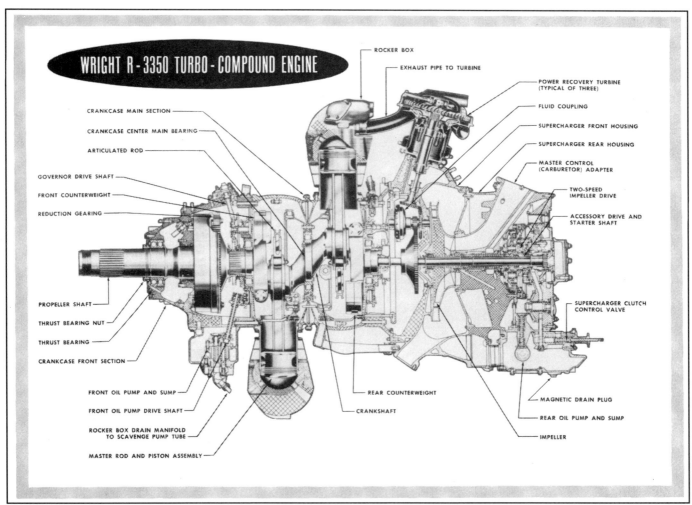

In the Wright R-3350 Turbo-Compound engine, the basic 18-cylinder engine has been supplemented with a second power producer, consisting of three "blow-down" turbines. (Boeing Historical Archive)

Original Airline Customers for DC-7 Aircraft

Airline	Quantity	Factory Serial Numbers
American Airlines	34	44122-44146
		45098-45106*
Delta Air lines	10	44261-44264
		44679-44684
National Airlines	4	44171-44174
United Airlines	57	44265-44289
		44903-44904
		45142-45156
		45356-45361
		45482-45490

* DC-7B structure but limited to DC-7 weight by use of DC-7 flaps.

TOTAL 105

Original Airline Customers for DC-7B Aircraft

Airline	Quantity	Factory Serial Numbers
American Airlines	24	44921-44925
		45232-45239
		45397-45407
Continental Air Lines	6	45192-45196
		45525
Delta Air Lines	11	44435
		45311-45314
		45350-45355
Eastern Air Lines	50	44852-44863
		45082-45089
		45330-45349
		45447-45456
National Airlines	4	45362-45365
Pan American World Airways	7	44864-44870
Pan American Grace Airways	6	44700-44704
		45244
South Africa Airways	4	44910-44912
		45477

TOTAL 112

brunt of the North Atlantic traffic until the introduction of the jet-powered Boeing 707 and Douglas DC-8 in 1959.

The DC-7C was also instrumental in the expansion of the over-the-North Pole routes by SAS; the Los Angeles/San Francisco-to-London service by Pan American World Airways; and the Amsterdam/Anchorage/Tokyo/Biak routes by KLM.

The DC-7C Seven Seas were the most desirable of the DC-7 series after major airline usage. Nonscheduled airlines such as Airlift and Saturn were quick to pick these aircraft up for their use in long-range charter, especially for the U.S. military. Others were converted to air tankers to fight fires.

Future Douglas transport production shifted to Long Beach, California, to take advantage of the available facility and required runway lengths to handle the jet-powered DC-8s. The facility at Santa Monica remained opened to handle some component fabrication, the modification of DC-6 and DC-7 aircraft to the freighter configuration, and some missile and space work. However as this work was shifted to new facilities in other nearby locations, the plant was closed and Douglas left the city of Santa Monica after spending over 50 years on or near the airfield.

DC-7CF

The DC-7CF Airfreighter was a DC-7C modified by the Douglas factory to allow the customer to configure the aircraft to carry either cargo or passengers.

As a transport, the DC-7CF could carry up to 38,000 pounds of cargo a distance of 3,000 miles, or 21,000 pounds more than 4,000 miles nonstop.

When passenger seats were installed in the convertible interior, it could carry up to 99 passengers and a crew of five nonstop across the Atlantic Ocean.

DC-7C(F)

A DC-7C converted to a freighter by a non-Douglas organization.

DC-7F

This was a DC-7B cargo conversion by Douglas Aircraft Company initially used by American Airlines.

DC-7 Airline Users

Only four U.S. trunk airlines placed the DC-7 into service, however others did operate the higher gross weight version DC-7B. American Airlines provided the impetus to production by placing the first order

Japan Air Line operated four DC-7C aircraft and was one of the last airlines to commit the "Seven Seas" to their Trans-Pacific routes. (Harry Gann Collection)

for 25 aircraft and subsequently placed the DC-7 into service between Los Angeles and New York on November 29, 1953.

Turboprop Proposals

Douglas undertook various studies looking at the application of turbo-prop propulsion for transports. One of the earliest was the model 1134C, which proposed a C-118A or R6D airframe with the Allison T38-8 engine developing 3,750 shaft horsepower. Douglas suggested that this aircraft could be used for cargo and personnel transport or as a test bed to study gust alleviation. A preliminary specification for this aircraft was issued in 1951. While the US Navy showed some initial interest, a satisfactory contract could not be negotiated and the study was cancelled.

Many of the studies for short- and medium-haul transports based on the DC-6 airframe were grouped under the Douglas Specification 1867 initially issued in early 1954. All of the 1867 series provided a new wing with a 15-percent thick Dou-

Original Airline Customers for DC-7C Aircraft

Airline	Quantity	Factory Serial Numbers
Alitalia	6	45228-45230
		45541-45542
British Overseas Airlines (BOAC)	10	45111-45120
Braniff Airways	7	45068-45074
Cia Mexicana de Aviacion	3	45127-45129
Japan Air Lines	4	45468-45471
Royal Dutch Airlines (KLM)	15	45180-45189
		45545-45549
Northwest Airlines	14	45203-45210
		45462-45467
Pan American World Airways	26	44873-44881
		44873-44887
		45090-45097
		45121
		45123
		45130
Panair do Brasil	4	44872
		45122
		45124-45125
Sabena	10	45157-45162
		45308-45310
		45495
Scandinavian Airlines System	14	44926-44933
		45211-45215
		45325
Swissair	5	45061-45062
		45190-45191
Compagnie De Transports Aeriens Intercontinentaux	3	45366-45367
		45446
TOTAL	121	

Douglas emphasized the speed in which the DC-7 aircraft could transport the public. Even as late as April 1955 (the date of this ad), Douglas could lay claim to transporting more people than "… all other airplanes combined." (Boeing Historical Archive)

glas designed airfoil (the normal wing was 16 percent). The model 1867G incorporated the DC-6B fuselage reduced in overall length by 60 inches (back to the standard DC-6 length) that accommodated 64 passengers in four-abreast seating. The powerplant intended was the Rolls Royce Dart engine, having 2,380 sea-level static equivalent shaft horsepower (eshp) for takeoff. It was designed for normal operations at ranges from 100 to 750 miles, with a maximum range of 1,000 miles.

The 1867G-1 was identical to the 1867G except that the intended power plant was the Allison T56-D10 derated to 2,500 eshp. The 1867H was to be equipped with the Allison T56-D8 rated at 4,050 eshp with a structure designed to exploit the speeds made possible by these larger engines. This version was designed to operate at ranges up to 2,000 miles. The 1867J was also to be equipped with the Allison T56-D8 engine, however it was to be a larger airplane with a 12.5-foot span extension at the wing center section with the DC-7C fuselage. It

This former PAA DC-7C has been converted to a freighter as shown by the replacement of the normal passenger door with the large cargo door, and was put into service with Overseas National as a DC-7CF. (Harry Gann)

Braniff Airways was the first domestic operator of the DC-7C, purchasing seven in 1955 for routes within the United States. Shown is the first aircraft in a flight out of Clover Field in Santa Monica. (Boeing Historical Archive)

Scandinavian Airline System (SAS) consists of an amalgamation of airlines from three countries, Denmark, Norway, and Sweden. LN-MOB is a Norwegian civil registration. (Boeing Historical Archive)

British Overseas Airlines, to remain competitive on the North Atlantic route, purchased 10 DC-7C aircraft when its Comets were grounded and the turbo-prop Britannia was delayed by design problems. After being delivered beginning in May 1954, BOAC disposed of their DC-7Cs in 1964. (Boeing Historical Archive)

Installation of the propeller hub encasing spinners along with cowl liners reduced drag while making more efficient use of engine cooling air. The wheelbase is slightly over 39 feet. (Boeing Historical Archive)

After Pan American and SAS confirmed their orders for DC-7C aircraft, Swissair followed with an order for two, later increasing it to three aircraft. Swissair was always a staunch Douglas customer, starting with the DC-2. (Boeing Historical Archive)

This over-water version of the DC-7C also had provisions for an aft sleeper compartment, as evidenced by the two small widows on this Sabena aircraft. This compartment could accommodate 16 passengers in seats or four passengers in seats and ten passengers in four upper and three lower berths. (Boeing Historical Archive)

Panair Do Brasil received four DC-7C aircraft including the first-built. Clearly seen is the five-foot wing center section extension, the 42-inch plug in the fuselage just forward of the wing leading edge and the two-foot taller vertical tail compared to the earlier DC-7 aircraft. (Boeing Historical Archive)

The prototype DC-7C Seven Seas is shown in an engine run-up. On the left wing tip is a flight test apparatus used to accurately record dynamic pressure. When compared to static pressure, the airspeed reading could be obtained. The vanes on the forward end of the boom were used to eliminate any effects of sideslip on the pressure reading. (Boeing Historical Archive)

N2281 was originally XA-LOB with Mexican. With President Airlines, it was named President Roosevelt. Later it was operated by Paramount Airlines (Harry Gann Collection)

would accommodate 82 passengers in a four-abreast coach interior.

One of the most serious of the Douglas DC-7C studies was the model 1848 and its various iterations. This was to be an over-water variant powered by the Rolls-Royce RB109 turbo-prop. The fuselage was to be increased by 40 inches over the standard DC-7C, and the new aircraft was known to the marketeers as the DC-7D. A similar proposal with Rolls-Royce Tyne engines was called the DC-7T.

None of these investigations into turboprop propulsion convinced Douglas that this was an effort that could be economically into production, despite the prodding of C.R. Smith at American Airlines. American subsequently bought the Lockheed 188, but as it turned out, that aircraft was not profitable as a commercial venture for Lockheed. Douglas had secretly started a project office in 1952 to evaluate the pure turbine engine possibilities, and the findings from this group convinced Douglas to bypass the production of aircraft using turbo-prop engines and concentrate on the production of transport that used the more efficient pure-jet engines.

In a memo that A.E. Raymond, vice-president engineering, wrote to Donald W. Douglas in late 1952, well before the final decision to not use turboprops in Douglas commercial transports, he stated that:

"We laid out the DC-8 with the idea that a turbo-prop version might sometime be built but whether that time will ever come depends on the date satisfactory engines are available and proven and the comparative advantages of jet and turbo-prop then.

Theoretically the turbo-prop has some advantages over the jet but it always lags behind the jet in development because it is more difficult and intricate design job due to the propellers and its controls, so this theoretical advantage is not counted upon and we will believe it when we see it."

A major contributing factor was the rapid development of the pure-jet engine. When Boeing, and later Douglas, made their 707 and DC-8 available, it hastened the end of the turbo-prop propulsion as a major player for all but the commuter airlines.

Time had finally caught up with the DC-4/DC-6/DC-7 concept as first-line transports. The use of turbine engines with a new total design to take advantage of the higher power and lower operating costs brought an end to the piston-engine era and relegated the turbo-prop powered aircraft to a small niche in aviation transportation. Douglas went to the four engine jet-powered DC-8 to compete.

The DC-7D was essentially a DC-7C with Rolls-Royce RB109 turbo-prop engines replacing the Wright R-3350 and the sweeping of the vertical tail. While various internal configurations were considered, this artist's rendering suggest a convertible option with the large fuselage cargo doors. (Boeing Historical Archive)

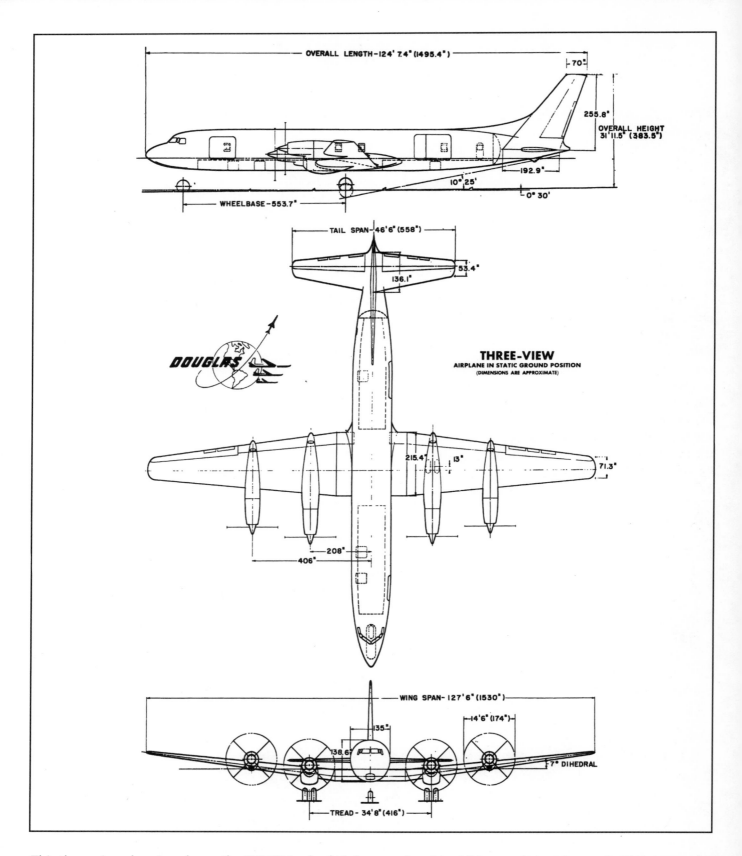

This three-view drawing shows the DC-7D in the freighter version. In addition to the engine and tail changes, the DC-7D also incorporated a 40-inch fuselage extension forward of the wing. British Overseas Airlines and American Air Lines were proponents for the use of turboprop engines (Boeing Historical Archive)

External Influence

The Lockheed Constellation — Chapter 6

The Lockheed Constellation in all of its variants has been described as the pacing transport in the immediate post-war expansion of the world's air transportation. Originally developed to compete with the Douglas DC-4, its initial production was sponsored by TWA and Pan American. The U.S. government chose to place the DC-4 with its higher reliability into production during World War II to meet long-range passenger and cargo needs.

When the original Constellation was released for commercial production, it could carry 60 passengers in a pressurized cabin versus 44 in the non-pressurized DC-4. TWA put the 049 Constellation into service on March 1, 1946, just 10 days after the first flight of the Douglas YC-112A. Western, National, American, and United Airlines began DC-4 service in early 1946. TWA and the Constellation clearly had the immediate postwar advantage, but it would not last long.

Although United and American Airlines had placed orders for the DC-6 prior to the end of World War II, Douglas was not allowed to begin civilian production until the course of the war had been determined. Most U.S. trunk airlines resumed service with converted C-54s. The DC-6 was finally placed into service in April 1947, eleven months after TWA's initial Constellation service. However, TWA lost its advantage when the Constellations were grounded from July 12 to September 20, 1946, after accidents caused by cabin pressurization supercharger problems. The DC-6 aircraft, in turn, were grounded from November 12, 1947 through March 21, 1948.

Lockheed worked fervently to eliminate the numerous reliability problems and to upgrade the capability of the 049 Constellation with the introduction of the 649 and 749 versions. Douglas countered with the DC-6B that entered service in 1951. Lockheed produced the stretched 1049 Constellation in 1951, followed by the improved 1049C in 1953. Douglas made the DC-7 available and it entered airline service in 1953, followed by the DC-7B in 1955. The 1049 series was then improved to the 1049G configuration and became available to the airlines in 1955.

When the DC-7C was introduced and Pan American put the Seven Seas into service in June 1956, the advantage clearly shifted to Douglas in the race for the top piston-engine transport. Lockheed countered with the 1649 Starliner, which entered service a year later. The Starliner had a greater range, but was somewhat slower. The controls, which were hydraulically boosted, were heavier to operate than the Douglas control system, which was aerodynamically aided by tabs. Only 44 Starliners

The Lockheed Constellation series was the pacing transport in the immediate postwar time period. In the end, Douglas overtook the Constellation and outsold it. An L-749 version is shown here. (Harry Gann Collection)

The airscoop located on the center under fuselage collected ram air for the cabin air heater. Fuel vented overboard and entering this intake was determined to be the cause of a fatal in-flight fire in 1947. The fix was to relocate the fuel vents aft of this airscoop to prevent the flow into it. (Boeing Historical Archive)

were built, while Douglas produced 121 DC-7C aircraft.

In the overall picture, Douglas built 2,283 of its four-engine DC-4/6/7 designs, while Lockheed produced 856 Constellations and Super Constellations. (These figures also included the military aircraft.) Clearly, Douglas won that race. However, in winning, Douglas lost sight of the possibilities that the upstart jet transports could have on the world air transportation market. Douglas struggled to keep abreast of a third company, Boeing, but in the end that battle was captured by the Seattle-based company. Douglas would never lead the field again and was eventually to disappear in 1997.

More widespread usage of the Constellation, especially during the early days, was hindered by the lack of development information to be shared with the world's airlines. Lockheed was left with a situation brought about the elusive Howard Hughes, who was the major sponsor with his controlling of TWA. His participation proved to be decisive as General Dynamics later found out with its 880 and 990 jet transport development.

The basic designs of the Constellation, while producing a graceful appearance, also introduced some major design problems during the stretching and capability improvements necessary to remain competitive. The triple tail was an attempt to ease entry through doors of smaller hangars then in use for maintenance. However, this approach had a detrimental effect on the flying characteristics with it destabilizing tendencies, especially on the shorter-fuselage Constellations.

Douglas had used the multi-tail configuration with one aircraft, the DC-4E, changing to a conventional empennage for the production versions. The graceful appearing curving of the Constellation fuselage also made it more vexing to stretch the fuselage, so necessary to meet the increasing capacity requirements as the world's airlines expanded to meet the rising markets.

MISCELLANEA

ACCIDENTS, FREIGHTERS, AND PYLON RACERS? CHAPTER 7

Although the DC-6 was introduced to airline service some 14 months after TWA had started Constellation service, that advantage was somewhat nullified when the triple tail Constellations were grounded by the CAB from July 12 to September 20, 1946, after accidents caused by the cabin pressurization superchargers. But American and United also faced a grounding problem when 97 DC-6s were set down between November 12, 1947, and March 21, 1948. This was precipitated by a fatal crash of a United DC-6 at Bryce Canyon, Utah, followed by an emergency landing at Gallup New Mexico, by an American DC-6.

Extensive investigation uncovered the fact that an in-flight fire could be started by gasoline which had overflowed while being transferred from the number 4 alternate fuel tank to the number 3 tank. When the crew failed to stop the transfer process in time to avoid overflowing the number 3 alternate tank, gasoline flowed through the number 3 alternate vent line, out the vent and, carried by the air slipstream, entered a cabin heater intake scoop under the fuselage. In both cited incidents, the cabin heaters were in operation. The gasoline entering the scoop ignited, causing the fuel to burn in the intake ducting and then to penetrate the air conditioning compartment. This combination of events resulted in a catastrophic fire in the case of the Bryce Canyon flight and a near-fatal event in the United incident at Gallop.

The changes to the DC-6 airframe to prevent further fires from this cause involved no major structural revisions. The alterations were to relocate the air intake scoops in the wing leading edge, while the overflow vents were place in the wing trailing edge aft of the cabin heater intake scoop. Fire extinguishers were relocated, and emergency flares were moved and later eliminated. After modifications were made, the DC-6 continued in service without any other major problems.

DC-7B/F-89J Mid-Air

On January 31, 1957, while on a routine production test flight, a DC-7B destined for delivery to Continental Air Lines collided with a Northrop F-89J Scorpion in the airspace of northern Los Angeles County. The DC-7B crashed into a junior high school yard, killing the crew of four and three students. The F-89J, which was on a functional

This Continental DC-7B was lost when a Northrop F-89J Scorpion collided with it during a production test flight. The DC-7B spiraled in with a locked hard over aileron deflection. (Harry Gann Collection)

Taxi accidents are always a problem on crowded ramps. Delta's first DC-7 took a hit from a DC-6 at the airline's headquarters at Atlanta. (Boeing Historical Archive)

check of a newly installed fire control system, crashed into a nearby canyon, killing the pilot.

The flight of the DC-7B was scheduled to be the last check flight, and was the first of five of this type on order for Continental. The airline was to introduce the aircraft to its flights between Los Angeles and Chicago on April 28, 1957.

Witnesses to the collision testified that the F-89 had completed a 90-degree turn when it struck the left wing of the DC-7B. The wing tip and part of the horizontal tail of the transport was stripped from the aircraft. Apparently as a result of the aileron jam, the DC-7 spiraled in.

No change was made to the design of the DC-7B, however a deposit was made to the Douglas design experience that surfaced on the DC-8. A cam was introduced into the DC-8 control system that would allow the crew, with a control force of 60 pounds applied, to move the opposite control surface, giving the pilot some control authority in case of a malfunction.

The CAB introduced new regulations prohibiting flight tests over metropolitan areas as a result of the incident. The use of chase aircraft or a knowledgeable observer was also required when the flight operations required more than normal preoccupation with cockpit duties.

In order to make the delivery schedule to Continental, Eastern Air Lines consented to give up one of its aircraft delivery slots. This replacement aircraft was allocated the same commercial registration, N8210H, as the destroyed airplane but received the next factory serial number, 45193. This aircraft was to end its flying career as a display at the Tallmantz Aircraft "Movieland of the Air" Museum at the Orange County airport in Santa Ana, California.

The forward lower cargo compartment on the DC-6B was sized at 267 cubic feet. On airplanes with four abreast seating, this compartment could accommodate 5,690 pounds of baggage or cargo. On aircraft with 5 abreast seating, the compartment was limited to 3,740 pounds. (Harry Gann Collection)

Development of Freighter Versions

On July 2, 1948, shortly after the inauguration of the DC-6 airline service, Douglas announced that "a huge, new modern air freighter" would be developed. Donald W. Douglas, president of the company further stated that:

> "The airplane will be the first commercial transport specifically designed, engineered and constructed for high-speed, low-cost, long-range transportation of air freight in the United States and over global trade routes. Already 60 per cent engineered, the prototype would be completed and flying next spring."

He went on to say that the greater speed and payload of the new air freighter, to be named the DC-6A, will enable it to do the work of two C-54's.

Development of the DC-6A was the company's recognition of the tremendous growth and potential of air cargo. The modest figure of 22 million ton-miles of cargo carried domestically by certificated airlines in 1945 expanded in 1946 to 90 million ton-miles. Entrance of non-certificated operators in the field and further expansion of freight operations by certificated airlines raised this figure in 1947 to 125 million ton-miles.

The DC-6A made its maiden flight on November 9, 1949, and remained with Douglas for flight test and promotional operations until delivered to Slick Airways on April 16, 1951, after receiving its type certificate.

As a means of extending the service life of the DC-6/DC-7 aircraft after the advent of the turbine-powered jetliners, Douglas offered a cargo conversion plan to the operators in 1958. This would convert these aircraft from the previous passenger to cargo carriers and would be a relatively cheap and fast supplement to the DC-6A to fulfill the immediate need of additional cargo lift caused by the rapid growth of the air cargo business. An obvious advantage of using the phased out piston-engined planes is the fact that they were well along toward being written off in depreciation.

The Douglas conversion offered varied options on the DC-6B and DC-7 variants, which began with the removal of interior passenger equipment and compartments, leaving a large unobstructed pressurized cargo compartment. If the customer so desired, some passenger-carrying capacity could be retained to provide a convertible capability. Optional was the retention of the passenger windows.

The Sabena Air Lines engineering group designed and converted two DC-6B to a swing-tail arrangement enabling 60-foot long objects to be carried. (Tim Williams via Harry Gann)

This DC-6BF was originally delivered to Pan American as a passenger carrier and was later converted to a freighter. The DC-6BF total volume capacity was almost 5,000 cubic feet and could transport 28,838 pounds 1,400 nautical miles on normal operating rules. (Tim Williams via Harry Gann)

The fuselage shell was strengthened to support the heavier payloads. Large fore and aft cargo doors and jambs were installed, as were new floor and floor beams. The cabin floor was fitted with either individual tie-down rings or flush lateral tracks which may be used for bulk or pallet loads. The fuselage were covered with a fiberglass-laminate lining.

American Airlines chose to convert 14 of its DC-7B passenger carriers to DC-7F Speedfreighters at a unit cost of $485,950 in a program that took Douglas about five months to complete. These aircraft could carry 35,800 pounds of mail and freight over 1,500 nautical miles and entered service in October 1959, supplementing American's 10 DC-6A all-cargo aircraft. The DC-6A could not provide transcontinental service at maximum payload without a fuel stop, whereas the DC-7F could. United Airlines also contracted with Douglas to convert six of its DC-7 aircraft for delivery beginning in February 1960.

The DC-7CF was a convertible DC-7C modified by Douglas and contained more than 5,000 cubic feet of cargo space. Conversion consisted of removing the normal passenger interior accommodations and installing large cargo doors, both forward and aft of the wings, and refitting a special floor stressed for the freight load. It could carry up to 38,000 pounds of cargo a distance of 3,000 miles or a 21,000-pound payload more than 4,000 miles nonstop. With seats installed, it could transport 99 passengers and a crew of 5 on a flight across the Atlantic Ocean without a refueling stop. Among the airlines opting for this modification were Alitalia, BOAC, Japan Air Lines, and Riddle Airlines.

By November 30, 1962, Douglas had reworked over 70 DC-6 and DC-7 aircraft for 15 different airlines. With the last of these conversions, Douglas shut the door of its Santa Monica facility to aircraft production after a 40-year run and the plant continued its space vehicles and missile fabrication for 13 years before leaving Clover Field for other locations.

In addition to Douglas, other companies also offered the freighter conversion service. There were many design options offered to prospective users such as swing tail, nose-loading, or turboprop power. Only the swing-tail option was taken up when the Sabena Airlines engineering department converted two from DC-6B aircraft in 1968. The DC-6B ST conversion made possible the transportation of products up to 60 feet in length. The cargo capacity of this conversion, which provided 2,591 cubic feet of usable space, could carry 25,353 pounds in the upper hold and 7,142 pounds in the lower compartment. One of these modified aircraft was delivered to Kar-Air, a Finnish charter airline. Spantax, a Spanish airline, operated the other DC-6B ST.

Howard Hughes

Dealing with the evasive Howard Hughes proved to be so time consuming that Mr. Douglas swore after doing some modifications on a Boeing biplane fighter type for Hughes that he would not do business with him again. That statement was perhaps given in the heat of protracted negotiations, for Hughes, through his Hughes Tool Company, did purchase two DC-6 series aircraft despite his closeness to the Lockheed Constellation with his control of the operations of TWA. They were a DC-6, factory serial number 43152, and a DC-6A, 45372.

On the DC-6, Howard Hughes

All of the Douglas DC-7BF freight conversions were equipped with DC-6A-size aft doors. They were 124x78 inches and were operated by the aircraft auxiliary hydraulic system. (Boeing Historical Archive)

insisted that the outer wing attach fittings be held to tighter tolerances than the usual Douglas production standards and even agreed to pay a premium for this feature. However, Hughes never flew the aircraft and before the airplane was delivered he sold it to Linee Aeree Italiane where it was passed on to Alitalia and then later to the Italian Air Force. This aircraft was built to Douglas Specification TS-1159, which was defined as an overwater, forward cabin day with an aft sleeper compartment.

The DC-6A was initially contracted for by Olympic but was taken over by Riddle, which then sold it to Hughes Tool Company, all of these negotiations occurring prior to Douglas delivery. Delivery consisted of attaching a truck to the towbar and rolling it across the field, to be kept grounded for the next 16 years. Thus it joined a growing number of aircraft that Hughes had stored around the country. After his passing, his company, Summa Corporation, took on the task of disposing of this phantom fleet. The DC-6A went to a nonprofit organization named the Partners of the Americas to fly mercy flights in Latin America.

Flying Television Transmitters

The versatility of the DC-6 airframe was proven in 1961 when two DC-6A aircraft were modified to transmit educational programs while flying at 23,000 feet altitude. Able to cover a 400-mile diameter receiving circle, the DC-6A could transmit lessons in science, history, mathematics, French, art, and music to students in 13,000 schools and colleges located in portions of Illinois, Indiana, Kentucky, Ohio, and Wisconsin.

Douglas and Westinghouse engineers designed and installed a 24-foot mast antenna that extended straight down from the bottom centerline of the aircraft. The design of the antenna permitted 360 degrees television transmission and was gyro stabilized in both pitch and roll axis to keep the antenna always pointed towards the center of the earth. This feature assured that the transmitting signal remained constant along the outer fringes with the aircraft in a pitch or bank attitude up to as much as 20 degrees. The system used to control the pitch axis of the antenna mast was also utilized to retract the mast to allow the aircraft to land. This Douglas designed and fabricated antenna was also equipped with de-icers that cracked the accumulated ice with cycled air pressure that inflated and deflated spanwise tubes.

While one aircraft was airborne, the other served as a backup in case of operational or transmittal difficulties. Purdue Aeronautics Corporation for the Midwest Program operated these two aircraft for several years for Airborne Television Instruction Company. They were originally delivered to Slick Airways and sold to Purdue in August of 1960.

Fighting Fires, Oil, and Bugs

The DC-6 and DC-7 played a major part in fighting large fires, controlling oil spills that penetrated bodies of water, and the continuous fight against insects that fight for the same space as the human population. They also have been used in the fertilization of plants and trees.

The ending of World War I, which produced an availability of cheap, surplus aircraft, also signaled the use of aircraft for fire spotting and fighting the spread of insects. The later use of the four-engine transports for carrying a quantity of fire retardant, oil dispersal agent, or insect repellent was facilitated by the passing of these aircraft from the scheduled airline's usage. The year 1971 saw the introduction of the DC-6 and later DC-7 series for this application.

This non-passenger/cargo practice has generated many companies that either manufacture the necessary plumbing and tankage or provided and operated the aircraft for this mission. Douglas Aircraft Company, as an adjunct to the flight test program on the DC-7, scheduled a test flight on December 12, 1953, using the prototype aircraft on its 84th test flight. Since the aircraft had a water ballast system incorporated in the flight test configuration to simulate various payloads, it was relatively simple to make a water drop on some ignited brush. This water ballast system consisted of six internal 400 gallon capacity tanks and appropriate plumbing, valves, and controls to permit discharging the water overboard in flight through three holes, 6 inch diameter each, in the bottom left hand side of the fuselage.

This demonstration was conducted at Rosamond Dry Lake, located eight miles northeast of Lancaster, California. The drops were made at altitudes from 150 to 600 feet, each dispersing from 300 to 500 gallons of water. The results were made known to the industry, and since the primary business of the Douglas Aircraft Company was to build transport aircraft, no further action was taken.

Most of the aircraft that have been operated in real firefighting operations used an external tank that was attached to the underside of the fuselage and faired into the mold line of the fuselage. It was found that this configuration also helped to direct the retardant down on the target rather than curl around the fuselage as sometimes occurred with internal tanks. DC-6 aircraft could carry up to 2,450 gallons of retardant, while the DC-6B and DC-7 series could carry 3,000 gallons.

When carried in an external, faired tank, the retardant had no tendency to circle around the fuselage and made a clean drop. DC-6B aircraft could carry 3000 U.S. gallons of liquid. (Harry Gann Collection)

In 1953, Douglas used the prototype DC-7 to demonstrate the feasibility of using aircraft to drop retardant on ground fires. Water, which was carried internally for ballast to simulate particular flight test points, was dropped on a ground-controlled fire. Without the use of external tanks the liquid tended to flow up around the fuselage rather than to drop vertically in a compact mass. (Harry Gann Collection)

On the aircraft used to spray insecticides, the tanks were usually internal with the spray bars located either over or under the wings. This arrangement was also use to deliver oil dispersal agents to combat oils spills. Fertilization of wild vegetation has been assisted by this delivery system to help stimulate the salmon population.

It was expected that with the availability of later turbine-powered military surplus aircraft that the Douglas transports would be phased out of this utilitarian operation. This has not occurred and DC-6s and DC-7s with the earlier DC-4s are still present in the world's fleet of air tankers.

Fatal Air Mission in Mauritania

In 1988, the U.S. Agency contracted T&G Aviation Inc., located in Chandler, Arizona, for the use and services of two DC-7C aircraft for international development for a locust spraying in Dakar, Senegal. After completing this, they were ferrying the aircraft to Agadir, Morocco. During this operation on December 8, 1988, both DC-7C aircraft were fired on and hit by shoulder-mounted surface to air missiles while flying at 11,000 feet altitude. N284, the lead aircraft was struck in the number four engine and, during the emergency descent, the wing failed at 5,000 feet altitude. All five of the crew were killed. The other DC-7C, N90804, was hit in the number one engine but was able to safely land at Sidi Ifni, Morocco. Polisario guerrillas, fighting against Morocco, admitted taking the action and apologized for their error in believing that the DC-7C were Moroccan military C-130 aircraft.

N284 was the first of 14 DC-7C aircraft ordered by Northwest Airlines and delivered February 28, 1957. It had been converted to a DC-7CF. N90804 was initially delivered to BOAC as G-AOIF on December 26, 1956. Several operators have used it.

Hurricane Hunters

One of the uses put to the DC-6 was that of a hurricane hunter. Two DC-6A aircraft were outfitted with special equipment, including a predominant radar antenna mounted to the lower forward fuselage. They were operated by the Weather Bureau of the Department of Commerce in the 1960s and later by NOAA when that organization took over the mission. Their mission was to penetrate a hurricane and record various data that would be available to help study how a hurricane functioned and to help develop an adequate warning system to reduce the massive wind and rain damage that occurs annually.

The two aircraft involved were DC-6A C/N's 45227 and 45368. Both aircraft were obtained from Trans Caribbean Air Lines.

Pylon Racing in a DC-7B

Among the design criteria that Douglas engineers did not take into consideration in the layout of the transports was that of closed-course racing against highly modified Warbirds such as P-51s, Sea Furies, Corsairs, P-38s, and Bearcats. Yet in 1970, this very thing did occur when Clay Lacy and Allan Paulson raced their DC-7B, an ex-American Air Lines transport, in the California 1,000-mile Air Race held at Mojave. This particular DC-7B had been converted to a freighter and after seeing service with American, was used by Zantop and Universal Airlines before passing to California Aeromotive, owned by Paulson.

The DC-7B averaged about 325 mph the closed course to finish sixth, using 4,100 gallons of fuel and 80 gallons of oil. Flying an average 60-degree bank, the aircraft experienced a maximum of 2.2 "Gs" in the turns. The race-area was laid out with 10 pylons on a 15.15-mile course and the race was for 66 laps, thus the race, which was billed as 1,000 miles, was really 540 miles.

In 1971 during a 1,000-mile, unlimited, closed course race near San Diego, California, the DC-7B made another appearance, this time joined by a Constellation. However, the pilots of the smaller plane objected to the possibility of flying through dangerous turbulence that would be generated by the four engine aircraft and the two transport aircraft did not compete. They did fly the course to allow the spectators to see the aircraft at a low altitude.

N66540C was one of two DC-6A aircraft used by the Weather Bureau of the Department of Commerce to track hurricanes, thereby gaining an insight into the manner in which these weather anomalies affect the general public. A search radar, which was attached to the underside of the forward fuselage, was used during the penetration of the fast-moving hurricanes. (Al Hansen via Harry Gann)

APPENDICES

APPENDIX 1 — DC-7 TRIVIA

On the Douglas DC-7 aircraft, some 638 suppliers provided raw and finished materials which contained:

- 16 miles of electrical wiring
- 4 miles of control cables
- 33,500 pounds of aluminum
- 350 pounds of titanium
- 1,007 pounds of plastics
- 1,485 pounds of fiberglass sheet
- 1,117 square feet of fiberglass cloth
- 95 instruments, including flight, engine, hydraulic, etc., of which 85 are in the cockpit
- The air-conditioning system could comfortably cool 48 average-sized living rooms on a hot summer day
- The heating generated by three heaters that provide the anti-icing could heat 25 average sized houses
- Basic blueprints to get the DC-7 into production would cover an area of 400 acres
- Electrical output of the DC-7 was 48,000 watts; sufficient to operate five average five room-houses.

APPENDIX 2 — DC-6 AND DC-7 RECORDS

DC-6

Date	City Pairs	Airline	Time	Remarks
October 4, 1948	LA – Jacksonville	Delta	6:43	320 MPH
June 3, 1947	LA – Tampa	National	6:50	354 MPH
June 3, 47	Tampa – Miami	National	0:39	312 MPH
October 23, 1948	Los Angeles – Atlanta	Delta	6:11	313 MPH
November 6, 1948	Los Angeles – Charleston	Delta	6:24	344 MPH
March 2, 1950	Chicago – Miami	Delta	3:08	376 MPH
December 3, 1950	LA – Mexico City	CMA	4:11	369 MPH

DC-6A

Date	City Pairs	Airline	Time	Remarks
May 1951	Philadelphia – Los Angeles	Slick		Largest single piece of freight
March 53	Los Angeles – New York City	Slick	7:45	
December 1, 1952	Los Angeles – Chicago	Slick	5:52	

DC-6B

Date	City Pairs	Airline	Time	Remarks
May 29, 1953	Los Angeles – Paris	TAI	20:28	
November 29, 1952	Los Angeles – Copenhagen	SAS	23:12	Over the North Pole
May, 1952	New York City – London	PAA		92 passengers
May 29, 1953	First scheduled world flight	SAS	83:4	Around the Globe

DC-7C

Date	City Pairs	Airline	Time	Remarks
November 17, 1956	Non-stop long distance	SAS	21:44	6,005 miles
May 23, 1957	Long Beach – Paris	KLM	21:35	6,146 miles

Summary of Production and Delivery Data — Appendix 3

Model	Quantity Sold	Authority To Proceed	First Flight	First Delivery	Last Delivery
DC-6	177 (1)	11/44	2/15/46	11/24/46	11/21/51
DC-6A/C	239 (2)	2/1/48	9/29/49	4/16/51	12/29/58
DC-6B	288	1/50	1/23/51	4/27/51	11/1758
DC-7	105	12/51	5/17/53	10/13/53	7/3/58
DC-7B	112	4/53	4/25/55	5/25/55	5/23/58
DC-7C	121	1954	12/20/55	4/20/56	11/4/58

(1) Includes 2 military
(2) Includes 165 military

Summary of Selling Prices — Appendix 4

Model	Base Price	Model	Base Price	Model	Base Price
DC-4	$ 534,945	DC-6A	$1,249,174	DC-7	$1.759,000
DC-6 (1)	$ 950,000	DC-6B	$1,120,000	DC-7B Overwater	$1,900,000
		DC-6B Overwater	$1,470,000	DC-7C Overwater	$2,272,000

(1) The first 100 DC-6 aircraft sold for $595,00 each

The clean design can be seen in this fly-by photograph of a United DC-6. (Boeing Historical Archive)

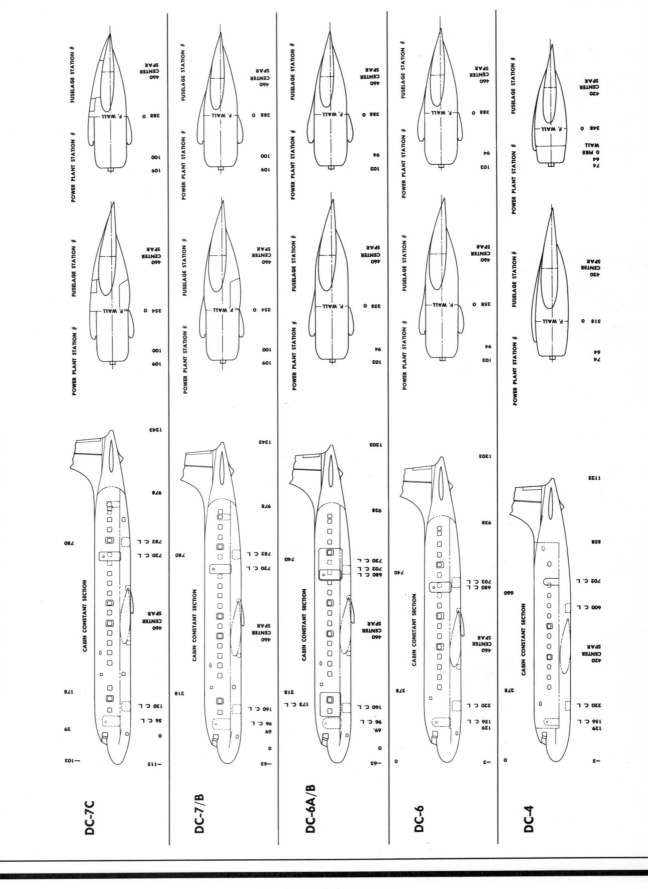

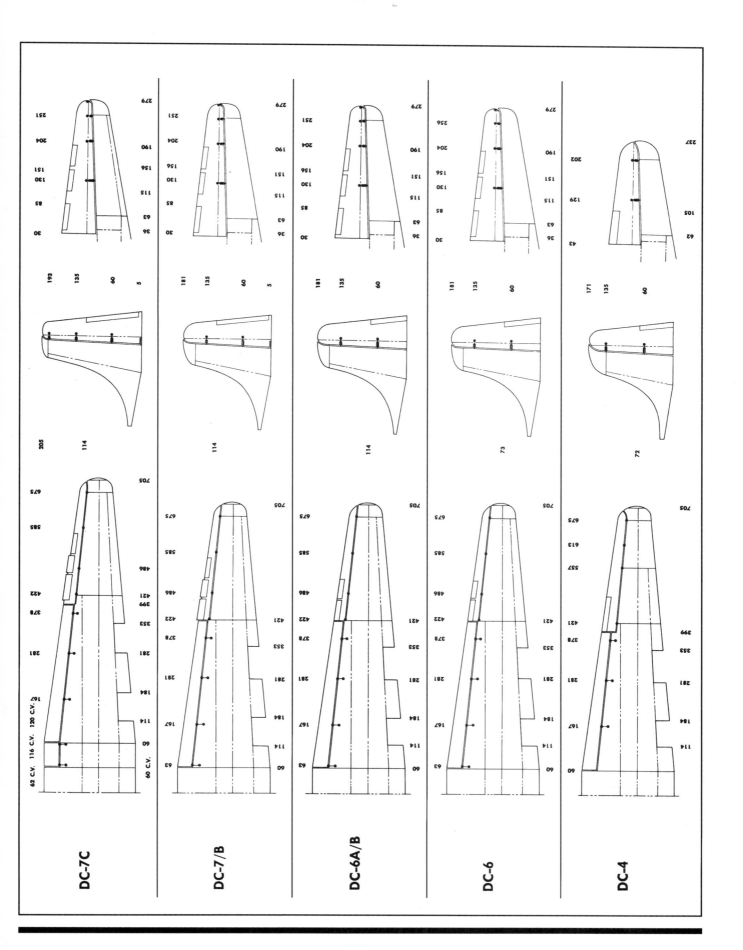

SIGNIFICANT DATES

IMPORTANT DATES IN THE HISTORY OF THE DC-6/7

July 22, 1920
Davis-Douglas Company formed

July, 1921
Douglas Aircraft Company formed

July 6, 1925
Douglas delivers its first transport: the M-1 to the U.S. Post Office

September 20, 1932
Douglas receives contract to build the DC-1

July 1, 1933
The DC-1 makes first flight

May 11, 1934
The DC-2 makes first flight

December 17, 1935
The DC-3 (DST) makes first flight

March 23, 1936
Agreement between airlines and Douglas to build the DC-4

June 7, 1938
The DC-4 (DC-4E) makes first flight

February 14, 1942
The DC-4 (C-54) makes first flight

July 1944
Douglas issues TS-477 for DC-6

October 23, 1944
Pan American announces contract for civil version of C-74 Globemaster

February 15, 1946
The XC-112 makes first flight

July 10, 1946
DC-6 makes first flight

November 24, 1946
First delivery of DC-6 to American and United Airlines

April 27, 1947
Both American and United Airlines begins service with DC-6

July 1, 1947
Independence (C/N 42881) was delivered as a Presidential transport

September 29, 1949
DC-6A makes first flight

February 10, 1951
DC-6B makes first flight

April 11, 1951
First delivery DC-6B

April 16, 1951
First delivery of DC-6A, to Slick

November 2, 1951
Last DC-6 delivered, (Braniff, fuselage 205, C/N 43295)

May 18, 1953
DC-7 makes first flight

June 3, 1953
First delivery of DC-6C (Sabena, fuselage 489, C/N 44421)

July 23, 1954
500th of the DC-6/DC-7 series delivered

December 20, 1955
DC-7C makes first flight

June 1, 1956
The "Seven Seas" enters service with Pan American World Airways

June 30, 1958
The 1,000th of the DC-6/DC-7 series aircraft is delivered

October 1, 1958
Last DC-6A delivered (Swissair, fuselage 1030, C/N 455510

October 17, 1958
Last DC-6B delivered (JAT, fuselage 1040, C/N 45564)

September 11, 1958
Last DC-6C delivery (Hunting Clan, fuselage 1025, C/N 45532)

December 10, 1958
Last of the DC-6/7 series is delivered (KLM DC-7C, fuselage 1041, C/N 45549)

November 30, 1962
Last aircraft rolls out of Santa Monica facility: a conversion of a DC-6B to cargo carrier

April 28, 1967
The McDonnell Aircraft Corporation and the Douglas Aircraft Company merge to form McDonnell Douglas Corporation

February 28, 1970
Last DC-6B in scheduled commercial operations makes final flight (UAL)

February 1, 1981
Donald W. Douglas Sr. dies

August 1, 1997
Douglas Aircraft, as a part of McDonnell Douglas Corporation, merges with Boeing. The Douglas name disappears.